HUNTING FOR STARS

THE McGRAW-HILL HORIZONS OF SCIENCE SERIES

The Science of Crystals, Françoise Balibar

Water, Paul Caro

Universal Constants in Physics, Gilles Cohen-Tannoudji

The Dawn of Meaning, Boris Cyrulnik

The Realm of Molecules, Raymond Daudel

The Power of Mathematics, Moshé Flato

The Gene Civilization, François Gros

Life in the Universe, Jean Heidmann

The Molecule and its Double, Jean Jacques

Our Changing Climate, Robert Kandel

The Language of the Cell, Claude Kordon

The Chemistry of Life, Martin Olomucki

The Future of the Sun, Jean-Claude Pecker

How the Brain Evolved, Alain Prochiantz

Our Expanding Universe, Evry Schatzman

The Message of Fossils, Pascal Tassy

Earthquake Prediction, Haroun Tazieff

MICHEL

MAURETTE

HUNTING FOR STARS

McGraw-Hill, Inc.

New York St. Louis San Francisco Auckland Bogotá
Caracas Lisbon London Madrid Mexico
Milan Montreal New Delhi Paris
San Juan São Paulo Singapore
Sydney Tokyo Toronto

English Language Edition

Translated by Isabel A. Leonard
in collaboration with
The Language Service, Inc.
Poughkeepsie, New York

Typography by AB Typesetting
Poughkeepsie, New York

Library of Congress Cataloging-in-Publication Data

Maurette, Michel.
[*Chasseurs d'étoiles*. English]
Hunting for Stars/Michel Maurette.
p. cm. — (The McGraw-Hill *HORIZONS OF SCIENCE* series)
Translation of: *Chasseurs d'étoiles*.
Includes bibliographical references.
ISBN 0-07-041029-1
1. Meteorites. 2. Astronomy — History. 3. Cosmology.
I. Title. II. Series.
QB755.M455 1993 93-23978
523.5'1 — dc20

The original French language edition of this book
was published as *Chasseurs d'étoiles*, copyright © 1993,
Hachette, Paris, France.
Questions de science series
Series editor, Dominique Lecourt

This book is printed on recycled, acid-free paper containing a minimum of 50% recycled de-inked fiber.

TABLE OF CONTENTS

Introduction by Dominique Lecourt 7

I. Meteorites and micrometeorites 19
The power of Jupiter 19
A first rendezvous on the Moon 27

II. The search, the stakes, and the heroes 41
Becoming a meteorite and micrometeorite hunter 41
A gallery of portraits of meteorite hunters 51
The hunt for micrometeorites 73
Still more meteorites? Why? 86

III. Approaching "primitiveness" 97
The usefulness of a classification 97
Desperately seeking parent 102
The passion for primitiveness 108
Hunting for isotopes 117
The microscopic hunt for presolar grains 129

IV. From the presolar time scale to the origin of life 135
The presolar time scale of cataclysms 135
The nebular time scale 142
Meteorites and formation of the earth 153
Meteorites and the origin of life 162
Meteorites and evolution of life 168

Conclusions 183

Bibliography 185

I was initiated into the science of meteorites and micrometeorites by three scientists who were most generous in sharing with me their knowledge of many years of practice: since 1984, Donald Brownlee (Department of Astronomy of Washington University in Seattle), the father of modern micrometeoriticists; since 1989, two of the world's most gifted mineralogists and meteorite specialists: Mireille Christophe Michel-Levy (University of Paris–VI) and Géro Kurat (Conservator of the collection of Austrian meteorites in Vienna), both of whom took a spontaneous interest in micrometeorites at a time when very few people did. Also, Ms. Christophe Michel-Levy graciously agreed to reread this manuscript and offer her highly constructive criticism. Finally, without my crucial meeting in 1962 with the American physicist Robert Walker; without the visionary intuition of the radio astronomer Jean-Louis Steinberg, who strongly supported in 1983 my first project to collect micrometeorites in Greenland; and without the decisive aid of the French Institute of Polar Research and Technology, I would not have written this book. To all I express my gratitude here.

M. M.

INTRODUCTION

This book is about meteorites and those who search for them. It presents the information gleaned in recent years on "micrometeorites," which are of such compelling interest that they have finally captured the attention of scientific circles. They are the very stuff of which stars are made, and Michel Maurette, who has an international reputation in the field of micrometeorites, works with astrophysicists, biochemists, and geologists to read the record of the solar system contained in these tiny bodies. But before investigators could focus on them, they had to overcome the obstacles raised by three thousand years of history.

The tales of the ancients tell us that stones have rained down from the heavens since earliest times. Egyptian papyri from the second millennium B.C. describe lumps of iron or stone striking the Earth. From such an occurrence that took place in Phrygia in 652 B.C. we have a precise description of what we call today a "meteorite": "A black stone in the shape of a cone with a circular base and an upper surface coming to a point." In the second book of his famous *Natural History*, Pliny the Elder (23 A.D.–79 A.D.) discovered the "torches" that flashed across the sky and remembered other points of light that behaved like fireballs, and sometimes: "... the heavens open, disgorging from the abyss matter resembling blood and fire which strikes the Earth." No greater terror for human beings could be imagined. Pliny credits Anaxagoras (500 B.C.–428 B.C.) with having scientifically predicted the fall of a stone from the Sun. We know from Harvey H. Nininger that in China, under the Han Dynasty (206 B.C.–220 A.D.), observations were made repeatedly of the same phenomena. Since the publication of work by Masako Shima, Sadao Murayama, *et al.* in 1983, we have learned that a stone which fell on May 19, 861 A.D. in Japan has been preserved. From Russia we also have descriptions of a "swarm of stones" falling in Vyshgorod in the second century and of the fall of a giant rock in 1296 near Velikiy Ustyug. A drawing shows the fall of a stone "the color of purple and as large as a sleigh" in 1584 near the Siberian town of Tashatkan, which owes its name to this event as it means "arrow of stone." In 1662, in Norye Ergi, also in Siberia, a small cloud was seen "from which fell bright, shining stones, some of them small, others large, all hot."

Anthropologists and historians tell us that humankind has traditionally attributed magical powers to these strange stones, as we see with the aborigines of Australia and the Amerindians. The Revelation to John makes several references to a fall of stars to Earth. It has been assumed, albeit without verification, that the famous black stone of the Kaaba in Mecca, a sacred stone venerated for many years before the birth of Islam, may in fact be a meteorite. This propensity to attribute a supernatural origin to these stones is splendidly illustrated by the fascinating and turbulent story of the most famous meteorite of all, of which a magnificent account was recently given by Ursula B. Marvin, a researcher at Harvard. This was the stone that fell near the town of Ensisheim in Alsace on November 7, 1492, a few minutes before noon. The sole witness to this fall was a young lad whom legend later turned into a shepherd boy. The sound it made as it crashed to Earth was heard in Lucerne, forty miles away. According to contemporary records, the impact of the stone made a hole—we would call it a "crater"—three feet deep, triangular in shape, with sharp angles. Its weight was estimated to be 300 pounds.

Before very long, crowds of people had chipped away pieces and carried them off for luck. The mayor of the town had it carried to the parvis in front of the church and it was later hung in the chancel, where it remained for three centuries. Now it happened that, on November 26 of that same year, Maximilian (1459–1519), the heir of the Hapsburg family and Archduke of Austria, who later succeeded his father, Frederick III, as Holy Roman Emperor, entered the town of Ensisheim at the age of thirty-three. Having heard of the stone, he caused it to be brought to him and decided that it was a miraculous sign, intended for him, announcing that he would soon win victory over the French, against whom he was preparing war. It was not long before poets and chroniclers took up the tale of the stone. The most celebrated of them, Sebastian Brant (1457–1521), a law professor at the nearby University of Basel, who had probably heard the crash himself, wrote two versions of the same poem relating the event, one in Latin and the other in German. A humanist and reader of Pliny, he tailored his description to the culture of the ancients but christianized it: the shape of the stone, which suggested a Greek delta, appeared to him to symbolize the Trinity, and so attributed a miraculous significance to the event. In the German vernacular version, he particularly emphasized the fact that this stone fell on the day of St. Florentin, a saint particularly venerated in Alsace. Perhaps not without a hidden political agenda, he suggested to his readers that this was a sign favorable to Maximilian. In 1960 it was discovered that this poem was the inspiration for a magnificent representation of the "miracle" of Ensisheim painted by Albrecht Dürer in 1494 on the back of the painting entitled the *Penitent St. Jerome*. Many a version of its prophetic significance was offered in the years that followed. First, Brant himself, in 1498, interpreted it as foreshadowing the death of Frederick III, who had cleared the path to the throne for Maximilian; it was also seen as encouraging a crusade against the Turks, called for by Maximilian himself in 1503. To accentuate the supernatural character of the event, the accounts soon shifted the time of the fall from noon to the dead of night.

Those that told of the stone crashing to Earth were in no doubt that it fell from the sky. But it took centuries for scientists to realize that such stones come from the stars. By a curious quirk of epistemology, modern science was more radically opposed to this recognition than Aristotle.

In his *De Caelo* [*On the Heavens*] and in the *Meteorologica*, Aristotle (384 B.C.–322 B.C.) provided an explanation of "shooting stars" according to the principles of his cosmology and physics. The bodies in the uppermost sphere of fixed stars are made of subtle, unalterable, and divine material (ether) which has no equal on Earth, in our "sublunar" world of generation and corruption. On the other hand, the "torches" and "goats" that we observe in the sky correspond to an "exhalation" phenomenon which is accounted for by the very structure of the cosmos. Just below the extreme edge of this uppermost sphere is a highly combustible "species of matter" which has both warm and dry properties. The flame is defined as the "ebullition of a dry exhalation" which catches fire when the circular motion of the sphere stirs this stuff up. "If the combustible material is broad and long," wrote Aristotle, "we often see a flame burning as in a field of stubble." If it is longer than it is broad, it seems to throw off sparks as it burns and is called a "goat" and when this does not happen, it is a "torch." The shooting stars appear when the whole length of the exhalation is scattered in small parts and in many directions.

"Sometimes," concludes the philosopher, "the cause of these shooting stars is the motion which ignites the exhalation. At other times, the air is condensed by cold and squeezes out and ejects the hot element, making the motion of these shooting stars look more like that of a thing thrown than like a running fire." Sometimes the motion of the star is in the manner of smoke exhaled by a lamp, and sometimes "the bodies are projected by being squeezed out like fruit stones from one's fingers and so fall into the sea and onto the dry land both by night and by day when the sky is clear."

Aristotle establishes an analogy and even a relationship (later confirmed) between this phenomenon and that of the thunderbolt, which is also ejected by pressure and thus hurled downward. He accounts for the oblique trajectory observed in both cases by the composition of two contrary movements: the natural movement of fire is upward while the "violent" expulsion caused by cooling carries it downward.

Complicated as it may appear today, this explanation governed scientific thinking until the very end of the 18th century. Even Descartes (1596–1650), the last person one would suspect of Aristotelian orthodoxy, defended in his seventeenth Discourse on *Meteors* (1637) a recast theory of exhalation, explaining that "a thunderbolt can sometimes be converted into a very hard stone which breaks and shatters all that it encounters if among the penetrating exhalations there are others which are heavy and sulphurous, especially if some are coarser, like the earth which settles to the bottom of a vessel of rainwater when it is left to stand." In any event, these notions imply that those stones raining down from the heavens have nothing to do with the stars and planets; this is why for a long time they were spoken of as "thunder stones."

At least those stones were believed to be extraterrestrial in origin. Paradoxically, the first effect of the dismantling of Aristotle's cosmology and of the discredit into which the Cartesian cosmology of the vortices fell was that this grain of truth became lost to scientists. The 18th century relegated them to folk tales and superstition. In 1771, Wolfgang von Goethe (1749–1832), who studied law in Strasbourg, made the trip to Ensisheim to examine this famous stone and showered sarcasm on the "credulity of the human species." He refused to see in it anything other than an ordinary stone.

A scientist as eminent as Antoine Laurent de Lavoisier (1743–1794), a mineralogist who was intensely interested in meteorology his entire life, when asked to examine a meteorite that fell in Lucé (France) on September 13, 1768, rejected the hypothesis of cosmic origin presented by Father Bacheley. In a paper submitted to the Academy of Sciences in 1772, he said that it was an "ordinary stone struck and altered by lightning which analysis showed to have no unusual properties." In similar fashion, Pierre Bertholon (1741–1800), editor of the *Journal des sciences utiles* [Journal of useful sciences], writing in 1791 about the meteorite observed on July 24, 1790, to fall in Barbotan, southern France, waxed indignant that such pure fantasy could be rationalized in an official report signed by the town fathers, considering that, as he stated, "it was a physically impossible phenomenon." Likewise, A. Stutz, speaking of an observation of the same type made in Hrašćina, Croatia, said that such folk tales had no place in the Age of Enlightenment. It is told that Thomas Jefferson (1743–1826) adopted the same attitude of radical skepticism in 1807 when a meteorite crashed to Earth in Weston, Connecticut.

And yet: by this time in history the first modern evidence of the extraterrestrial origin of these stones had been supplied by Ernst Florens Chladni (1756–1827), a Russian of German origin and a corresponding member of the Academy of Sciences of St. Petersburg, in a short book published in Riga in 1794. Chladni's arguments were based on the examination of two very large stones, one weighing 680 kg [1500 lb] ("Pallas") found in Krasnoyarsk, Siberia, and the other, an enormous mass of iron weighing 13,500 kg [29,700 lb] discovered in Otumpa in the Gran Chaco area between Paraguay and Argentina. He based his analysis on the equal distribution of olivine in the body of the Pallas meteorite and the absolutely "exotic" nature of the iron of which the Otumpa meteorite was composed, even though it was found in a schistose region with not a trace of minerals. He concluded that these objects had indeed fallen from the sky.

Obviously he attributed no supernatural cause to the phenomenon. He rejected the explanation that these meteorites were artificial or the product of an accidental conflagration. They are, he said, masses both of which were hurled from a very far-off region. He eliminated the volcanic hypothesis, believing that pure iron could not have been ejected by volcanoes, and noted that, after all, neither of the two regions in question had ever had any iron.

Studying the trajectories of shooting stars, he concluded that they must be solid bodies falling into the atmosphere under the effect of gravitation. He finally suggested that all these objects (meteors, shooting stars, stones, and metal erratic boulders) must bear some relationship to one another.

Although in actual fact twenty-five years elapsed before Chladni's hypothesis took hold, it was reinforced by the observation of numerous falls that happily occurred just after the publication of his book: a hail of stones in Sienna, Italy, starting on June 16, 1794; a large stone the following year in England that terrified a farmer in Scarborough, Yorkshire; and another which fell in Krakhut, India in December 1798. The latter cast into terminal doubt the theory of condensation that equated meteorites with thunderbolts: the Krakhut meteor fell from a clear blue sky.

But it took progress in chemical analysis before it was possible for anyone to accept the extraterrestrial origin of these objects without being dismissed as a medieval "obscurantist."

In 1800, Charles Barthold, a professor of chemistry at the Ecole Centrale du Haut-Rhin, removed a large sample of the Ensisheim stone taken from the church authorities by the revolutionaries and placed in the custody of the Colmar Public Library. He subjected it to chemical analysis but arrived at the mistaken conclusion that it could have been torn out of a nearby mountain!

At about the same time, the young English chemist Edward C. Howard (1774–1816), at the request of Sir Joseph Banks (1743–1820), president of the Royal Society, was conducting analyses of the same type and coming up with very different results. Was this because the "natural theology" of the British saw no antagonism between science and religion and entertained less antipathy for anything that fell from the sky? In any event, with the assistance of a French mineralogist, Jacques-Louis, Count of Bournon, who had left France under the Reign of Terror, he arrived at some decisive results.

The two scientists observed the presence of "curious spheres" on these stones (today we call them "chondrules") and noted the presence of nickel in the metal grains they contained. All stones of this type of which they had some knowledge were of similar composition, and all were different from stones found on Earth. The two concluded that they absolutely had to have come from the sky.

When he read Howard's paper, the French chemist Antoine-François de Fourcroy (1755–1809) had an 11-kg [24-lb] sample of what remained of the unfortunate Ensisheim stone brought to him. Nine kilograms [20 lb] of this sample are still in the Paris Museum of Natural History today. The results of his analyses, published in the fall of 1803, agreed with those of Howard and with those obtained at the same time by Nicholas Louis Vauquelin (1763–1829).

Pierre Simon de Laplace (1749–1827), Jean-Baptiste Biot (1774–1862), and Siméon Denis Poisson (1781–1840) agreed. They hypothesized that these stones came from volcanic eruptions on the Moon. This was the hypothesis that prevailed until the 1840s, but at least the science of meteorites was able to make a start, at the crossroads of astronomy, chemistry, geology, and physics. The lunar origin theory was replaced by a scenario connecting meteorites to the nebular hypothesis of the formation of the solar system as put forth by Immanuel Kant (1724–1804) in his *Theory of the Heavens* (1755) and then, based on different information, by Pierre Simon de Laplace.

Today it is the antiquity of meteorites that interests researchers, who use them to test hypotheses on the origin of our solar system. They may hold the key to the question of the origin of life: through the channel of biochemistry, molecular biology is joining astrophysics in the new branch of knowledge known as "exobiology." Since the seminal work of Luis and Walter Alvarez published in 1979, meteorites have contributed to turning paleontology upside down and thrown many an expert into consternation! Could they throw light on the catastrophe at the end of the Cretaceous 65 million years ago, the chief result of which was the disappearance of the dinosaurs?

The reader can look forward to a thrilling story whose heroes had a sense of mission. The expeditions to the North Pole by the American Robert Peary (1856–1920) at the beginning of the century, the doggedly persevering research of Daniel Moreau Barringer (1860–1929) and the strenuous quest of Harvey H. Nininger

(1887–1986) have become history. Our contemporaries have taken up the challenge—from William A. Cassidy to Michel Maurette. But in the meantime new objects even more promising for understanding the most distant past have captured our attention: the tiny "micrometeorites" to which the author of the present book has devoted his professional life. Of necessity, this researcher has become a hunter searching for stars, and, as we shall see, his profession is by no means a reposeful one. From Greenland to the Sahara, these "hunters of stars" have emerged as adventurers in the truest sense of the word—and their adventures have included fighting off smugglers. The administrative maze they had to negotiate when they returned to their laboratories proved to be no less a mine field. This is a vision of scientific activity which in many ways breaks with its stuffy official image. Maurette restores its full dimension of risk, which, after all, is its human value.

Dominique LECOURT

I

METEORITES AND MICROMETEORITES

THE POWER OF JUPITER

Our solar system, formed from the Sun and the nine planets* accompanied by their sixty-one satellites, is also populated with a huge number of far smaller and lesser-known objects in the enormous family of "small bodies," which are the "stars" of this book. Unlike the planets, these small bodies sometimes collide with the Earth and deposit on it a very ancient material. This material contains unique information about our origins and may also be the cause of cataclysmic destruction. Among the largest of these bodies are all the comets, over a thousand billion of them, that populate the Oort's cloud, a region that extends about 50,000 astronomical units** around the Sun; to this we must add the thousands of asteroids that orbit mainly between Jupiter and Mars. Despite their large number, the total mass of the comets represents only

* Four terrestrial or inner planets: Mercury, Venus, Earth, Mars; four giant planets: Jupiter, Saturn, Uranus, Neptune; and the last planet, Pluto.

**The astronomical unit (AU) is the distance between the Earth and the Sun, about 150 million km [93 million mi].

ten times that of the Earth and that of the asteroids, 5% of the mass of the Moon, which is in turn one-eightieth the mass of our planet. These comets and asteroids are the "parent" bodies of innumerable smaller objects (between a few hundred meters and a few ten thousandths of a millimeter in size): the meteoroids and micrometeoroids. When such objects less than a few yards in size fall to Earth, they produce a "shooting star" in the atmosphere. The materials they leave behind in the atmosphere are called *micrometeorites* if they are less than a millimeter in size and *meteorites* if they are more than a few centimeters. Between these sizes (roughly 1 mm, and 1 cm) are the "minimeteorites," for which we will be searching in Greenland in 1995.*

These tiny bodies may be as crumbly and porous as snow; they may also be very hard, compact, brittle, or malleable. Some of them have a very complex structure: a single micrometeorite one-tenth of a millimeter in diameter may prove to be made of a porous aggregate composed of millions of individual grains included in an organic matrix. Some, however, have a simpler structure: they consist of a single silicate crystal belonging to varieties that abound in extraterrestrial material as well as on Earth, such as olivine and the pyroxenes.

The smallest of them may resemble tiny grains of cosmic sand comparable in size to viruses which can be

* These minimeteorites are very difficult to find because they are below the limit of detection with the naked eye as one walks through a hot or cold desert, but are not plentiful enough to be extracted from the Antarctic ice like the micrometeorites.

seen only under a very powerful microscope. At the other end of the scale there are enormous comets: cosmic "Everests" made of dirty ice about 10 km [6 mi] in diameter, and stony or even iron asteroids like super-Everests. The largest of these monsters, Ceres, would come close to covering about half the area of Texas.

These objects orbit at dizzying speeds, between 11 and 72 km/sec [7 to 44 mi/sec], while a shell fired from a cannon moves at only 5 km/sec [3 mi/sec]! We are still unable to simulate these hypervelocities on Earth except in the case of micrometeoroids, which can be fired out of a dust gun, a sort of giant compressed-air pistol. Those meteorites no bigger than a few meters that travel on appropriate trajectories intersecting that of the Earth can be captured by our planet; when this happens they are sharply decelerated by the Earth's atmosphere, which causes them to heat up to the point of partial destruction. They then form a kind of fireball (called a "meteor") accompanied by various acoustic phenomena which vary according to the mass, velocity, and friability (crumbliness) of these bodies.

Every year the Earth collects about 20,000 tons of micrometeorites and about a hundred tons of meteorites. In fact, this amount of material is quite small. In the course of four billion years, assuming a constant rate of deposition, a uniform layer only 20 cm [8 in.] thick would have been deposited on the surface of the Earth! For comparison, during a period of "quiescent" volcanic activity (no major eruptions), the upper layers of the atmosphere at altitudes between 20 and 30 km [12 and 18 mi] (the "stratosphere") contain about 300,000 tons of very fine volcanic products. This quantity goes up to a few million tons as soon as there is a major eruption.

When they are as large as ten meters [30 ft] or so across, and too heavy to be slowed by the atmosphere, meteorites explode on impact with the land or ocean. Only a few fragments that fly off in the atmosphere can make it to the ground without being destroyed. A body 10 km [6 mi] across self-destructs, giving off the energy of a million hydrogen bombs striking the ground at the cosmic velocity of 20 km/sec [12 mi/sec] since such a large object is not slowed down by the atmosphere at all.

Most of these objects turn out to be far older than the rocks of the Earth or the Moon, and it is mainly their "antiquity" that makes them so interesting. In fact, the majority of them were formed at the very beginning of the solar system, at the same time as the Sun, from an interstellar presolar cloud of gas and dust usually called the "solar nebula." In this nebula, most of the interstellar grains were partially metamorphosed. Those that melted completely and then suddenly cooled, or even those that were volatilized and then recondensed, following events that are still mysterious to us, gave rise to a new generation of grains defined as having been "formed in the solar nebula." This mixture of grains was then incorporated into "parent" bodies, the asteroids and comets, where they remained trapped for billions of years. Then, "recently,"* when such a parent body collided with another body, these grains were relaunched into space in the form of meteorites and micrometeorites, some of which were "fired" in the direction of the Earth.

It is estimated that the meteorites were formed from these primitive grains between 4.5 and 4.6 billion years ago. The materials of which the Earth and the Moon are made also have ancient origins but were modified into rocks, the oldest of which were formed between 3.8 and 4.2 billion years ago, respectively. Because of this, the material of the Earth and Moon lost its "primitiveness": these two bodies were sufficiently massive for internal energy sources to heat them to the point that their constituent material was melted over and over again so that the original properties were lost.

A high point in the science of meteorites was a battle started in 1960 by three United States meteorite scholars (Robert E. Fish, Gordon E. Goles, and Edward Anders), the effects of which have rippled down to the present day. Until 1960, scientists believed that the main classes of meteorites came from a small, somewhat Earth-like planet which had melted. They assumed that it had an onion-like structure with an iron-nickel core, followed by a mixture of iron and rocks in contact with the core, topped by a mantle of rock, all of which was covered by a surface crust. To develop such a structure, the little planet must have been strongly heated, so that meteorites

* "Recently," on the geological scale, means up to about 100 million years ago.

could not be primitive. If this were so, they would be of no more scientific interest than samples of the Earth, Moon, or Mars. The hypothesis of the three American scientists changed the entire picture: they declared that meteorites were derived from numerous parent bodies sufficiently small to have preserved their primitiveness. This hypothesis, which has since been verified time and time again, contributed to a renewal of interest in meteorite research, turning it into a very lively discipline. Moreover, the launching of the first Sputnik in 1957 and the efforts that followed revived curiosity in everything that came from space.

Today we can say that asteroids are bodies that failed to accrete into a planet. For reasons that remain mysterious, they never grew large enough to be the size of small planets where the temperature would have been sufficiently high to melt them down. The absence of such a planet, which would have been located between Mars and Jupiter, allows us to collect primitive meteorites today and "read" from them essential information on the earliest moments of the solar system. Without this "miraculous" gap, astrophysicists could never have gone as far as they have in their calculations of "nucleosynthesis" (to be explained below), which are based on very precise measurements of the abundance of elements in the most primitive meteorites.

This situation should probably be attributed to the rate with which Jupiter was formed, shortly before the other planets and the asteroids. Jupiter is the most massive planet in the solar system, with a mass that is 320 times that of the Earth. Its gravitational field could have been sufficiently strong in the asteroid belt to capture or eject most of the mass scattered in the form of small bodies, thus preventing them from accreting into a planet.

So, meteorites and micrometeorites appear to be the oldest cosmic sediments we have and the only ones that allow us a glimpse back into the first stumbling moments of the solar system. Some of them even contain a small portion (less than 0.1%) of tiny *presolar* grains (one thousandth to one hundred thousandth of a millimeter), older than the Sun, which survived the primitive evolution of the solar nebula. We will see how scientists have learned to use them *one by one*, actually doing the work of archivists of our presolar history.

In fact, this presolar matter—interstellar gas and dust, born of gigantic stellar cataclysms—is what we are all made of. After numerous metamorphoses, some of them violent, this matter generated not only Space Station Earth which sweeps us around the Sun at nearly 30 km/sec [18 mi/sec], but also the calcium of our bones, the iron in our red blood cells, the 70% water of which our bodies are made, the fluorine in our toothpaste, and so on.

This little book will tell the story of how the scientists known as "meteoriticists" met the meteorites and micrometeorites, our mineral ancestors. This meeting first takes place on collecting expeditions, then in the laboratory, where we try to decipher some of the most interesting *fossil messages*. Springing from the imagination of the researchers is a scenario that, down the cosmic time scale, allows us to describe the main events that mark the presolar history of the constituent grains of meteorites, through explosions and eruptions of stars, then analyze their turbulent primitive history in the solar nebula at the time the Sun was born. We "see" them swirling in something like an immense and highly dilute cirrus cloud that takes the form of a blizzard near Jupiter and beyond. Titanic "electromagnetic" thunderstorms roared through them, snuffing out and generating particle after particle. But the role of meteorites and micrometeorites does not stop at this earlier epoch, but extends into the formation and evolution of the Earth, and particularly the origin and evolution of life, which began about 4 billion years ago.

We are still groping our way through this mist, this mass of laboratory analyses, theoretical calculations, and astronomical observations used by scientists to create these scenarios. Happily, the mist is gradually clearing (very gradually) thanks to a kind of *collective spirit* among scientists, the fruit of collaboration between astronomers and astrophysicists—who study the solar system from the "outside," trying to pierce the mystery in the gas and dust cocoons where the stars form—and planetologists of all disciplines, who look at the solar system from the "inside," analyzing extraterrestrial material in the laboratory or looking out at it from Earth through appropriate telescopes or instruments aboard satellites.

We call all the bodies found in space or on Earth "meteorites," and often conveniently forget the distinction between meteorites and meteoroids and sometimes even between meteorites and micrometeorites, according to widespread usage in most scientific articles. Our first stop will be on the Moon with the United States astronauts who flew seven Apollo missions to our own satellite (six of which were successful) between July 19, 1969, and September 12, 1972, since these missions gave a decisive impetus to our researches.

A FIRST RENDEZVOUS ON THE MOON

The Earth possesses in its Moon a huge natural detector that records impact craters produced by the smallest bodies of the solar system. This detector is unique by virtue of its area, about 38 million km^2 [14.7 million mi^2], and the time for which it has been exposed to the meteorite flux (over 4 billion years), especially its light zones —the lunar "highlands," which are older than the darker-colored "seas" or "maria." The micrometeorite detectors carried aboard satellites since 1960 are rarely more than a square meter [10 ft^2] in size and have not been exposed for more than a few years.

Unlike the Earth, the Moon has been dead for at least 3 billion years. No volcanic eruptions or continental drift have buried the craters. Since the Moon lacks an atmosphere, neither rain nor wind has altered their shapes; the incessant bombardment of micrometeorites and meteorites is the principal source of erosion. But this is an extremely slow process: it will take about 100 million years, for example, to erase the first human footprint on the Moon! What is more, meteorites are neither slowed down nor destroyed by passing through an atmosphere as on Earth: thus, they strike the surface of the Moon directly with their full mass and at full cosmic velocity. The geometric shapes of the craters tell us something about the characteristics of the impinging bodies and give us otherwise inaccessible information on the flux of large meteorites that affect the Earth-Moon system; of course, we are studying in particular its variations in the past. These studies have profoundly altered our ideas about the early evolution of the Earth.

The American astronauts of the Apollo mission were the heroes of a highly organized hunt for lunar samples. First, they explored the maria, the dark zones that can be seen from the Earth with the naked eye in bright moonlight. By exploring six different lunar sites, they were able to bring back a total of nearly 380 kg [836 lb] of material that included both rocks and samples of the sand which covers the entire lunar surface to an average depth of 5 m [16 ft]. In fact, the rocks and grains of sand brought back to Earth can be equated with lunar meteorites and micrometeorites. These collecting expeditions were supplemented by the work of the Soviet Luna 16, 20, and 24 probes, the first "hunter" robots designed to collect extraterrestrial samples and bring them "automatically" back to Earth. The meteorite hunters operating on Earth have added to this collection the twelve lunar meteorites (i.e., chunks of the Moon) discovered since 1976 on our planet.

Before the 1969 lunar landing, scientists disagreed as to the origin of the lunar maria, the largest of which, Mare Imbrium, is about 1500 km [930 mi] in diameter, which is nearly half the Moon's diameter. Some believed that we were seeing the traces of ancient seas while others imagined that they were volcanic craters. Still others held them to be enormous craters due to the impact of giant bodies with hypervelocities greater than 10 km/sec [6 mi/sec]. Since the first historic human steps on the Mare Tranquillitatis, the question has been settled: they were caused by the impact of gigantic bodies.

Lunar discoveries followed fast on the heels of each other. Who can forget the picture of the first footprint on the Moon, looking just like the footprint a person would leave in wet sand on Earth? But there is no water on the Moon! We concluded that the surface of the Moon was not made of rock alone. After the enormous impacts, the craters became filled with molten lava flows which, as they solidified, turned into basalts similar to those that formed the ocean floors on Earth. This solidified lava was then subjected to a rain of bodies of all sizes. Slowly, the cosmic rain wore down the lava flows into sand—very slowly, because this ground-up residue known as "regolith," or mantle rock, builds by only one meter [3 ft] every billion years. So direct observation shows the surface of the Moon to be riddled with impact craters of every size by bodies which "dig into" the floor of the oceans to an average depth of about 5 m [16 ft].

To investigate how the number of craters varies as a function of their size, we avoid observations in areas "saturated" with craters. When we focus on the microcraters, of which there are far more than the large craters, we carefully choose in the lunar soil superb balls of glass a few millimeters in diameter. These are droplets of lunar rock or soil that melted when struck by meteorites, splashed upward, and solidified on their way back to the lunar surface. Initially, they were as smooth as glass on the outside, but exposure to the rain of meteorites riddled them with hundreds of microcraters measuring between one-thousandth of a millimeter and one millimeter. It has been observed that the number of craters drops off sharply as their size increases.

When we look at the larger craters, we see that there are slightly over 300,000 of them over 1.6 km [2.6 mi] across on the visible side of the Moon but only 87 giant craters over 100 km [62 mi] across in its entire area, although this is over 100 trillion times the area of a glass ball! We derive the following law from adding up the crater areas: the total area of the 100 million 10-m [33-ft] craters is equivalent to that of the 10,000 1-km craters and to that of the few rare craters over 100 km [62 mi] in diameter observed in the maria. This shows that there is no discontinuity in the size of small bodies in the solar system. They simply become rarer as their size increases. Although we observe some discontinuities on Earth, these must be attributed to the destructive impact of meteorites striking the Earth's atmosphere.

Through laboratory simulations which consist, for example, of firing projectiles at layers of sand, we can establish a ratio between the diameter and depth of a crater and the mass of a projectile. When we study the size distribution of the lunar craters, we can determine their mass distribution and then calculate the "flux" of these projectiles to the Moon. When we evaluate this flux on Earth,* we must also take into account *gravitational focusing* which doubles this flux. This effect is linked to the ability of one body to exert gravitational attraction on another.

Thus we know that, in the vicinity of the Earth, just before entry into the atmosphere, we find about one micrometeorite measuring one-tenth of a millimeter per square meter per year, and that bodies measuring five-tenths of a millimeter are 250 times less abundant. Thus, the main distribution of micrometeorite mass is in the form of particles a hundred microns** in diameter. Conventional meteorites are far fewer in number since,

* In the most recent evaluation of meteorite flux, we combine studies on lunar impact craters with studies on impacts of the tiniest micrometeorites with detectors aboard satellites, and the local phenomena (meteors) that the astronomers observe in the visible and radar wave ranges from which they can deduce the mass and velocity of the projectiles.

**A micron is one-thousandth of a millimeter; hence a 100-μm micrometeorite is one-tenth of a millimeter.

according to current estimates, about 30,000 meteorites weighing over 100 g [3.5 oz] fall on the *entire Earth, per year.* In the same surface area, 550 million km^2 [212 mi^2], larger bodies are rarer still: the impact of an object with a diameter of 10 m [33 ft] would occur every hundred years, one with a 100-m [330-ft] diameter every 10,000 years, and a few with diameters of 1 km [0.6 mi] every million years. As for the truly cataclysmic impacts of objects over 10 km [6 mi] in diameter, only one would occur every 20 to 40 million years.

This sharp decrease in the number of meteorites with increasing mass appears to be a specific characteristic of the meteorite flux. It is not found, for example, when a hard rock is crushed in the laboratory and the masses of fragments obtained are measured. When this is done, a far higher proportion of large fragments is found than in the meteorite stream.

A lucky thing too, for us here on Earth! If such bodies had struck the Moon, even larger bodies would have had to impact the Earth. Since they are extremely rare, the probability of intercepting them increases in proportion to the area of the collector and its ability to exert gravitational attraction on another body, which is in direct proportion to its mass.

Thus a hypervelocity impact (15 km/sec [9 mi/sec]) of a body measuring 10 m [33 ft] across (the size of a small building) already produces an explosion with a power equivalent to the first atomic bomb dropped on Hiroshima (about that of 20,000 tons of dynamite). With a body 100 m [330 ft] in diameter, the equivalent of a thousand bombs of this power at the same time at the same site would explode releasing a total energy of 20 megatons (millions of tons). The first estimates that could be made for the Mare Imbrium indicate that its parent impact would have been produced by a body about 100 km [62 mi] in diameter. Given a speed of 20 km/sec [12 mi/sec], what an explosion! It would have released a trillion times the energy of the Hiroshima bomb. Such an explosion was almost powerful enough to split the Moon up into several fragments; on Earth it could have wiped France off the map, thus spilling the Atlantic into the Mediterranean.

When these first observations and calculations were made, people should have thought about the incredible "nuclear" winters that would be triggered by such impacts on Earth. But it was not until twenty years later that this idea took hold, when a group of scientists around Luis Alvarez suggested that the extinction of the dinosaurs could have resulted when an object (*one-tenth the size* of that believed to have formed the Mare Imbrium) struck the Earth.

Exploration of the Moon led to another important discovery. It had long been observed that the light-colored lunar highlands had far more craters than the maria, which are much darker. Through measurements made on rocks brought back by the astronauts from six zones of different ages, we now know that this increase in crater density is in proportion to the age of the collecting surface. Since the lunar highlands are several hundred million years older than the maria, we can prove that the meteorite flux increases considerably as we go back in time. Four billion years ago, for example, when the first life forms appeared on Earth, this stream was probably a thousand times more intense than it is today.

This increase, which has been called *accretionary tail*, shows that the solar system was "dirtier" back then. It was cluttered with the accretion debris that gave birth to the planets and their satellites; but in less than 500 million years these residues were eliminated, either by falling onto the planets or their satellites or being hurled out of the solar system by the "gravitational slingshot" of the planets, particularly Jupiter—the "swing-by" effect that we use today to place space probes in specific orbits.

Our rendezvous with the Moon also resulted in the forging of a new community of scientists who applied new working methods. Since the lunar samples were highly valuable, NASA immediately organized research teams where several groups studied very small quantities of samples together; this habit of teamwork has continued to the present day when we study the rarest meteorites and micrometeorites. Until 1980, NASA strongly encouraged us to present new results every year at the Houston Lunar and Planetary Science Conference, which has been held regularly in mid-March since 1970. With no new results, there was little chance of renewing the "principal investigator" contract which gave access to lunar samples and research funding. Today, the most dynamic groups in the study of meteorites and micrometeorites never fail to show up at the annual Houston "examination," where it is advisable to show how brilliant you are.

The preparation for the conquest of the Moon also led to the development of highly sophisticated techniques of microanalysis for analyzing very small samples of material. With considerable foresight, NASA has always supported their use since about 1960. One of the most powerful instruments applied to lunar samples and other samples of extraterrestrial material is the *electron microprobe*, developed in 1951 by the Frenchman Raymond Castaing, at the time a student of Professor André Guinier, who suggested that he build it as part of his thesis. This extraordinary instrument analyzes the chemical composition of tiny quantities of material: no more than 1 cm^3 divided by one trillion.

This microanalysis race is still run by groups that have mastered the use of a Castaing microprobe coupled to an electron microscope and that have been fascinated by the exceptional potential of another microanalytical instrument, the *ion microprobe* developed by a student of Castaing, George Slodzian. This instrument, which is so complex that it is still difficult to exploit all its potential, turned micrometeorite studies upside down in 1985. It can be said that this school of French physicists literally revolutionized our working methods; in any event, I myself would never have joined the micrometeorite search without them, as they let us extract a maximum amount of information despite the extremely small size of the samples.

The same techniques, once they have been automated, will in the 21st century allow us to analyze samples from comets and Mars returned to Earth by the descendants of our first hunter robots. This automatic analysis will allow us to study a very large number of grains or detect extremely rare mineral phases. The instruments will operate day and night, controlled by a sort of preprogrammed highly computerized robot.

Determination of the flux of meteorites according to their masses, the demonstration of the accretion tail, the motivation to use and improve new techniques of microanalysis, and the emergence of new multidisciplinary meteoriticists are all part of the most interesting scientific spin-offs of the walk on the Moon. One might object that these few spin-offs from the lunar program, which have proved so useful to the expansion of our discipline, appear to be quite futile to society at large in view of the very high cost of this enterprise (approximately $30 billion). How can knowledge of the micrometeorite flux and accretionary tail serve us? How much has lunar science cost? A little background information will be useful in answering these questions, which are often raised as objections.

The meteorite flux can be compared to a sort of heavy machine gun hidden in space lying in wait for astronauts. A "large" micrometeorite, one millimeter in diameter, moving at a speed of 20 km/sec [12 mi/sec] is capable of punching a hole 2 cm [just under an inch] in diameter and 1 cm [nearly half an inch] deep in aluminum. A "small" meteorite 1 cm across, traveling at the same speed, would produce an enormous hole: over 20 cm [8 in.]. Before sending people into space for very long periods, we would do well to ensure that the flux of micrometeorites and meteorites was sufficiently small not to depressurize a spacecraft by suddenly piercing it.

As to the Everests several kilometers across, their explosive power is equivalent to that of terrible atomic bombs, far more powerful than any device fabricated by humans. Because we have determined the lunar flux, we know there is no point in hurrying to leave the solar system, because these monsters are very rare. But there is some point in developing telescopic early warning methods for timely detection of incursions of such bodies into the close "suburbs" of the Earth and in thinking about ways to deflect them before they hit our planet.

The advantage of being familiar with the accretionary tail which existed over 4 billion years ago seems to be no less evident. Indeed, it has radically changed our ideas about the early evolution of Earth. Moreover, according to a new "exobiology" scenario, this tail could be responsible for the origin of life on Earth. Is it important to understand such issues? If not, is it important to understand anything at all? Why, for example, would scientists continue to investigate new insects even when we already know about 750,000 varieties? Why would they pursue the no less innumerable varieties of fossils? Why do mathematicians now study the geometry of chaos; and so on? Without this broth of new observations and new hypotheses, without constantly testing our knowledge and ceaselessly searching for truth, there would simply be no further progress of scientific knowledge.

Let us not forget that the primary goal of the lunar walk was not scientific at all, but a decision made by the United States government to take up the challenge of the first Sputnik launched into space by the Soviet Union in July 1957. We owe a great deal to those who we now call familiarly our Four Musketeers: the Americans Jim Arnold, Paul Gast, Robert Walker, and Jerry Wasserburg, who toiled in the shadows for years so that the lunar walk would become a scientific walk as well. Indeed, researchers, as always, have been used as a pretext to justify the cost of the conquest of the Moon. But they stand falsely accused: it is not their 382 kg [840 lb] of samples that cost $150 billion! They made sure that these fabulous sums would also bring back something for science. They should be acknowledged for this achievement.

The cost, evaluated for the decade 1960–1970 as being less than one fourth what Americans spent on cigarettes, is actually comparable to that of a major weapons program, such as the B-2 stealth bomber. We don't want Northop and their subcontractors to continue laying off workers, but we might wonder whether a program aimed at walking on the ocean floor or exploiting the mineral resources of the Moon or an asteroid wouldn't be a better source of new jobs and technologically useful discoveries. As yet, present-day society is not able to set up peaceful projects in place of weapons programs designed to kill, probably because the latter have the enormous advantage of becoming rapidly obsolete. Hence the endless arms race that does not necessarily have a good cost/job ratio.

Actually, this ratio turns out to be far better in the case of the Moon program because another aspect, rarely mentioned, of this adventure is that about 500,000 people worked together for ten years to send human beings to the Earth's satellite with maximum "reliability." This is an absolute record because the Manhattan Project set up during the Second World War to develop the atomic bomb ahead of Germany "mobilized" so to speak no more than 100,000 men and women.

In addition to the inevitable military application, this giant collective effort had innumerable technological spin-offs which have flowed imperceptibly into our everyday life. I will mention only three to illustrate their diversity. First of all, the development of computer science, over 50% of which is estimated to be due to this progress, then the development of "remote medicine" techniques using probes derived from those used to monitor the health of astronauts in space; and finally transmission of their health reports by satellite. Thus, a doctor working, for example, in one of the Peace Corps-type organizations in a developing country and faced with a difficult diagnosis can now use a direct satellite link to get help from a specialist in a hospital who instantaneously sends back the results from specialized equipment. Another spin-off particularly useful in the

United States in these days of ruthless technological competition is the development of new management techniques which have enabled 500,000 people from very different backgrounds, particularly researchers in industry and at universities scattered throughout the land, who do not meet as a matter of course, to work together for the first time on the same objective.

Such revolutions in work habits have always been the source of substantial technical progress. It may be that historians of the future will recognize this change as being an important step in human history, on a par with the mastery of fire or the invention of the steam engine.

II

THE SEARCH, THE STAKES, AND THE HEROES

BECOMING A METEORITE AND MICROMETEORITE HUNTER

Because micrometeorites are not hunted in the same way as meteorites, we will consider the two ways of hunting separately. First we will look at conventional meteorites with masses of more than 100 g [3.5 oz], namely objects a few centimeters [about an inch] in size that can be held in the hand. For the hunter, there are two types of meteorites: falls and finds. While the former are seen to fall and are picked up as quickly as possible, the finds may have been exposed to weathering for an appreciable period of time. Researchers prefer falls because they are "fresh": there has been no time for them to be corroded by natural water, microorganisms, or roots, as often occurs with finds collected in temperate latitudes.

However, researchers have about as much chance of collecting a fall themselves as they do of winning the lottery. This can be shown by a simple calculation. Every year, about 30,000 meteorites weighing over 100 g [3.5 oz] fall over the entire surface of the Earth, three-quarters of which is covered by water. So no more than 10,000 meteorites per year will land on the continents, with their 5 billion inhabitants.

Suppose your back yard has an area of 200 m^2 [2,150 ft^2] and you keep it under constant surveillance with an array of video cameras or seismographs. You would have to wait twenty million years before just one meteorite would fall right into your vegetable garden! This explains why there are so few accidents due to meteorites falling to Earth. The count for the last hundred years was one dog killed, one person slightly injured, and one meteorite that went through the roof of a house and landed on the living room sofa, on which happily no one was sitting at the time. The most recent incident was the night of October 9, 1992, when a 13-kg [29-lb] meteorite pierced the trunk of the car of an eighteen-year-old woman in the New York suburbs.*

Your chances of picking up a "find" are far greater, provided you choose your site carefully. Ideally, you will look in a desert area where the life expectancy of meteorites is far longer because there is less weathering by water and microorganisms. To increase your chances, you should zero in on areas favored by various phenomena.

---

* This was an ordinary stony meteorite, of no great value. The young woman who owned the car, who was no fool, offered the object for ten times the price ($15 a gram) to dealers. History will remember her asking price for her car: $10,000, for a car worth no more than $900 before the accident. Well, it was the first cosmic traffic accident ...

The best hunting grounds are certain blue ice fields in Antarctica where meteorites have frozen at the surface of the ice and survived for hundreds of thousands of years. Currently, the record is 900,000 years for stony meteorites! In hot deserts this life expectancy is about a few tens of thousands of years, but in temperate climates it drops to a few hundred years.

Logistically, however, Antarctica is not exactly the easiest place to get to: very costly, government-funded expeditions have to be organized, one has to camp under rough conditions hundreds of miles from the coast, and run substantial risks. To date, no travel agency has organized a meteorite safari in Antarctica.

If you insist on doing your own hunting, it is easier to go to a hot desert like the Sahara. One of my ideas, although so far unsuccessful, has been to ask the people who run Club Med to organize one of their adventure safaris in the Sahara to hunt for meteorites; of course, they would have to work things out with the local authorities so as to avoid being arrested as smugglers and forced to spend a few years in prison. For $3,000, insurance included, one could hunt meteorites for two weeks under the expert guidance of a meteoriticist who would recount the history of the solar system under the stars at night.

However, these hot deserts are not as kind to meteorites as Antarctica because they are not as dry. Contrary to appearances, the air above a desert contains more water vapor than at the poles, where residual water vapor is rapidly condensed out by the cold. Moreover, hot deserts undergo major climate fluctuations. We know, for example, that the Sahara was far wetter 10,000 or 20,000 years ago. Rain fell often, and its surface was studded by innumerable lakes and rivers. The life expectancy of meteorites would be no more than a few tens of thousands of years, except in the Great Eastern Erg, which was far drier. Here, meteorites are rare, estimated to be about 10 per km^2 [4 per mi^2] per million years. If we bear in mind that the life expectancy of a meteorite rarely exceeds 50,000 years, we would have to search through about 1 km^2 to find just one.

This is not impossible if you use a four-wheel-drive vehicle and have one person looking out each side at a strip of land 10 to 20 m [33-66 ft] wide. Driving 3-6 mph for a month of eight-hour days, you can pick up thirty or so in an area where they have accumulated for about 50,000 years. This is about the score of today's smugglers (whom we shall discuss later) at the top of their form. In two months, driving an ordinary 4WD vehicle, they manage to collect about a hundred in a good year. There are even areas more favorable to accumulation where the pickings are better.

The trickiest part, in hot or cold deserts, is to choose the right spot. In the hot deserts, we should not repeat the error committed in 1990 by some of my eminent meteoriticist colleagues who set out to explore an area containing iron-rich stones. These colleagues did not have the experience of guides who lead tourists through the desert, nor the experience of those scientific specialists in the study of past desert climates who, as part of a big UNESCO program, traveled over 100,000 km [62,000 mi] in a 4WD.

During the first telephone call I made to them as part of my preparation for a meteorite-hunting expedition in the Sahara, these guides and specialists began by confidently assuring me that they had never seen as much as a single meteorite. As I was familiar with smuggling exploits, I insisted on going to see them and brought along some photos of these smuggled meteorites. I showed them the first photograph accompanied by the brief comment: "It's this type of stone, covered with a sort of black varnish, that interests me." After thinking for no more than ten seconds, these people with very different backgrounds all gave me the same answer — either a groan or an oath: they certainly had seen quantities of meteorites in the course of their peregrinations, but had thought they were ordinary terrestrial basalts. Then they added: "Above all, you shouldn't get into volcanic formations or formations containing iron ore because your meteorites will look like stones from these types of rocks when you try to spot them from a slowly moving vehicle."

The picture I showed them was that of a splendid black stone resting on a background of light sand. To a specialist, there was no mistaking it. The stone is black because, as it fell through the atmosphere, it was heated by friction with the air molecules which melted its surface layers and, in most cases, caused it to lose over 90% of its initial mass. The final residue of this fusion is a surface crust at most 1 mm thick that is dark and glassy if the meteorite has freshly fallen to Earth, but gradually loses its gloss with time as it lies out under the stars.

There is another point: if you turn over the black glossy stone in your hand you may well find a fracture surface in which the black varnish was cracked, revealing a lighter-colored internal section. This contrast between the dark, glossy surface and the lighter-colored interior will be your first identifying criterion. Then, if the only instrument at hand is a magnifying glass, you will see on the fracture surface of most of these dark stones small round objects ranging from a few tenths of a millimeter to a few millimeters in size. These grains do not occur in terrestrial rocks; they are called "chondrules" (from the Greek *chondrion*, or "little grain"). They are very characteristic of the most abundant stony meteorites which for this reason are designated "chondrites."

But if you search only for stones with a black gloss, you will miss a lot of meteorites because this gloss fades with time. In fact, the first rule of the teams of scientists who search for meteorites in the blue ice fields bordering transantarctic mountains and cluttered with moraine debris* consists in becoming very familiar with local terrestrial rocks. Once the eye is trained, meteorites can readily be distinguished, and the diagnosis will almost always be confirmed by subsequent analysis. The same rule is applied in hot deserts.

---

* That is, local fragments which became detached by the grinding action of a glacier as it flowed on its bed rocks to the sea.

Another group of easily identifiable meteorites comprises the iron meteorites, as their constituent iron-nickel alloy confers on them properties comparable to those of steel. An iron meteorite is heavier than a stony meteorite of the same size because iron has twice the density of stone. The surface is generally furrowed and pitted with characteristic dents as large as a thumb, due to flows of molten iron as the meteorite fell through the atmosphere.

Why did our desert specialists earnestly recommend that we not venture into an area near a volcanic massif like the Hoggar or near a zone containing rocks with a high iron content like certain regions of Mauritania? Because any stone lying in a desert will slowly weather until it is finally covered by a darker patina. So it will soon look like a meteorite if it was already dark to begin with. This "desert patina" is attributed to microorganisms which digest the top millimeter of rock surfaces so that they have the same appearance. Although its genesis is still poorly understood, it is well known to remote sensing experts who attempt to assess desert mineral resources from photographs taken from surveillance satellites. As the patina masks resources deeper down, all the geologic formations of a desert end up looking alike. This *similarity of patinas* prevents meteorites from being found in desert areas rich in basalts or in rocks with a high iron content. When looking for them from a four-wheel-drive vehicle, you stop, you leave no stone unturned, and not a single meteorite do you find.

Smugglers, who have to be productive, prefer not to operate in temperate zones because they know that the life expectancy of stony meteorites is no more than a few hundred years. After this time they "fossilize" like ordinary pieces of wood and undergo a *terrestrialization* which makes them unrecognizable. What is more, in these regions there are far too many dark terrestrial stones that are indistinguishable from meteorites. So smugglers operate in light-colored desert areas, far from volcanoes and iron ore deposits.

The problems encountered in collecting micrometeorites are altogether different since they are huge in number compared to meteorites: over 10 quadrillion per year for the whole Earth, instead of 30,000. If you walk for half an hour at moderate speed the clothes you wear may well catch one! But you might as well look for a needle in a haystack, because, as it falls, the micrometeorite is lost in a huge mass of terrestrial particles that whirl around us all the time, whipped up by winds and other air currents.

You might then climb to the top of Mt. Rainier, away from the hiking trails, to make the most of the pure air swept by high-altitude winds. You then melt a few tons of snow to collect a hundred grams [3.5 oz] of sediment containing about a hundred million grains, but among them you will find only a few micrometeorites one-tenth of a millimeter in diameter. A lot of trouble for very little! Besides, they also corrode very rapidly under the influence of acid rain; their lifetime is probably no more than a few months. So if you were able to wave a magic wand and find a few, they would be useless for scientific study because they would have lost their essential "primitive character" due to rapid transformation of their component minerals into ordinary little grains of terrestrial clay.

This is why you have to go to the polar ice of Greenland and Antarctica, where micrometeorites are protected by freezing at low temperatures. The ice, moreover, is ultra-pure, nearly as pure as doubly-distilled water. Its contamination with terrestrial grains is less than one ten-thousandth that of the snows on Mt. Rainier. When you melt the ice and filter the meltwater through screens, you can collect sediment very rich in micrometeorites.

The extreme rarity of terrestrial sediment in Antarctic ice obviously facilitates extraction of micrometeorites, which have a concentration in the range of 5 to at most 200 per ton of ice in the most favorable zones. So you have to melt a lot of ice in Antarctica, involving logistics beyond the range of the ordinary tourist. In Greenland, which is far more accessible, the situation is different because the Arctic summer is warmer. Enormous quantities of ice melt near the margin of the glacial ice cap in the "melt" zone which extends over a few tens of kilometers. This natural meltwater flows down a multitude of rivers that crisscross the ice, sometimes rushing at a speed of over 3 m per second [10 ft/sec]. In the calm meanders of clear water pooling on the blue ice, you are surprised, in this desert of shining ice, to see patches of black mud. This is the “cryoconite” where we have been able to discover high concentrations of micrometeorites.

Extracting them from the sediments released when Antarctic or Greenland ice melts is relatively simple. As with the larger meteorites, you choose the darker grains, those with no bright colors and which look like little grains of coal. Almost always, in-depth analysis will show that you struck it lucky. The dark color is accounted for by the fact that most micrometeorites are initially made of a material that resembles the rare and most primitive, dark-colored meteorites; moreover, as they fall through the atmosphere, they become coated with a very thin layer (less than a thousandth of a millimeter) of a black iron oxide called “magnetite.”

As we have seen, it is not easy to hunt down the “finds.” Today, the difficulties inherent in the natural environment are compounded by sociopolitical obstacles. For example, the present-day rebellions of the Tuareg in Africa have led to highly mobile roving bands of looters who will not hesitate to attack and grab everything you have, sometimes leaving you stark naked out in the desert, where the nights are cold. Even if a peace treaty is signed between the Tuaregs and the government, it is far from certain that this looting will come to an end.

Despite these difficulties, thousands of people are excited enough about meteorites to devote their lives to them.

A GALLERY OF PORTRAITS OF METEORITE HUNTERS

At the present time there are about five thousand ardent meteorite collectors in the world. From a long list, I have chosen a few portraits of these hunters of extraterrestrial material to illustrate the variety of their personalities and motivations. We will see the daring risks they have been taking since the 1870s to collect meteorites and micrometeorites and we will see the strokes of luck and ingenuity that brought them success in their expeditions.

PEARY AND THE ESKIMOS

After the Scottish polar explorer John Ross met a group of Inuit (Eskimos) in 1818 near Cape York in Melville Bay, 150 km [93 mi] south of Thule in northwest Greenland, American, British, and Danish expeditions followed for about seventy-five years with the hope of finding one of the large meteorites mentioned in the folklore of these people named "Iron Mountain." But the Inuit stubbornly refused to disclose its location.

In May 1894, the American arctic explorer Robert Peary, glowing from his exploit of crossing the Greenland ice cap on dog sleds, managed to convince a hunter to lead him to the famous "Iron Mountain" by slipping him a few gifts. After two days of bad weather the hunter gave up and was replaced by another, named Tellakoteah. He, too, threatened to quit but finally, after a great deal of hesitation, he revealed to Peary the existence of three great blocks of iron (*Saviksue*, or "Great Iron"),. which in local legend constituted the mystic trilogy of the Woman, the Tent, and the Dog, driven from the heavens by Tornasuk, the spirit of evil. Since time immemorial, the Inuit had been using the "Woman" as the single source of iron for making the tips of their harpoons and forming the cutting edges of their bone knives. So they made judicious use of the chips of this extraterrestrial iron which we now know to have properties similar to those of steel.

After traveling for more than ten days across an obstacle course, Tellakoteah dug through the snow until he revealed the "Woman" (3000 kg or 6600 lb). Peary returned to Cape York the following year to collect the "Woman" and the "Dog" (410 kg or 900 lb) with no great difficulty. Then in 1896 he returned with his new ship, reinforced as an ice breaker. He moored it to a sort of natural jetty a hundred or so meters [330 ft] from the enormous "Tent" (2.5 m tall by 1.5 m wide [5 × 8 ft], nearly 30 tons of iron and nickel). After ten days of effort to move the meteorite with hydraulic jacks and before succeeding in hoisting it on board, he had to flee Cape York to avoid being trapped by the ice which was closing in sooner than expected. It was another three years before he made a fresh attempt, but this time he succeeded: after a tussle of over ten days with the enormous mass of the "monster" which the Inuit believed to be inhabited by a demon, despite bad weather and icebergs which were getting dangerously close, he managed to load the meteorite onto the deck of the ship by pulling it onto a pontoon made with three-ton tree trunks reinforced with railroad rails. The meteorite was christened "Ahnighito" with a bottle of wine and on October 2, 1897, Peary sailed into Brooklyn. It took a team of eighty horses to pull the meteorite through the streets of New York and bring it to the Museum of Natural History, where it has occupied a place of honor ever since (in the Hayden Planetarium) beside the "Woman" and the "Dog." Today, there is not a nation in the world, including Greenland, that would allow such exports, not to say extortions, of meteorites.

The weight of the "Tent" (about 31 tons) was not determined until 1956. The price for which he sold it (about $50,000) allowed Peary to finance his expedition to the North Pole, which he claimed to have reached on April 16, 1909. This claim remains controversial. In 1926, and again in 1964, two Danish expeditions returned to Cape York and brought back two other meteorites, one of which, weighing 20 tons, was christened the "Man." All these meteorites belong to what is called a "multiple fall," resulting from fragmentation in flight of one large iron meteorite. It appears probable that other chunks of this multiple meteorite fall may still be hidden in the Cape York region.

THE BARRINGER CRATER

We will now leave Greenland for a superb desert of Arizona, near the city of Flagstaff. Here we find the best-preserved and the most famous impact crater. It looks like a circular amphitheater 1,300 m [about 4,250 ft] in diameter and 170 m [558 ft] deep. Today we know that the projectile was an enormous iron meteorite weighing about 70,000 tons, the size of a small building. Its residues constitute the group of the so-called Canyon Diablo iron meteorites. A few tons of them have been collected. It is variously known as "Canyon Diablo Crater," "Meteor Crater," or "Barringer Meteorite Crater." The latter name was proposed by the official Meteoritical Society in honor of Daniel Moreau Barringer.

Barringer was an amazing man: trained as a lawyer, he switched careers to become a prospecting geologist; this brought him to South America, Spain, and finally Arizona, where he discovered a rich silver mine, the earnings from which afforded him a comfortable income. His first steps in the Meteor Crater convinced him, contrary to prevailing scientific opinion, that it was the impact crater of an enormous meteorite that had the same composition as the iron meteorites found near the crater. He estimated its mass to be 15 million tons. He wagered that most of this mass would be found buried at the bottom of the crater. Since, in addition to iron, the meteorite contained useful metals such as nickel, cobalt, and even platinum, he invested his own fortune as well as capital from enthusiastic shareholders to undertake drilling projects starting in 1904 to find the buried meteorite.

The project took nearly twenty years and none of the drills, some of which went down to 420 m [1,378 ft], struck the meteorite. Many bore holes did penetrate water tables, however, so that the site was frequently flooded and work was held up more than once.

Doubting that the crater was caused by an impact of this kind, scientists made a laughingstock of Barringer and even the most optimistic shareholders backed out. In 1929, just before the Depression, Barringer hired a theorist, F. R. Moulton, to determine the size of the meteorite more precisely in order to reassure the shareholders. (Moulton had served as an explosives expert in World War I when he designed shells.) The results of the calculations were catastrophic: Moulton showed that the energy released upon the supposed impact would have been sufficient to evaporate most of the meteorite. Instead of the 15 million tons initially predicted, the impact residue could be less than 5,000 tons! The last investors then abandoned Barringer for good. A few months after the Moulton report came the Crash of 1929, and Barringer died shortly thereafter.

The great merit of Daniel Moreau Barringer was that he compelled the conservative scientific community of the time to recognize grudgingly the existence of the first impact crater on Earth. He had invested in the adventure not only his personal fortune but close to thirty years of his life. The decision of the Meteoritical Society to change the name of the crater to Barringer Meteorite Crater in 1950 appears completely justified. Today, with far more sophisticated calculations using observations from artificial explosion craters, the mass of the projectile is estimated at about 70,000 tons and its speed at 55,000 km per hour [34,000 mph]. The volume of the meteorite can be compared to that of a small five-story building. These calculations have also confirmed that the main mass of the meteorite would have volatilized instantly on impact. Only small fragments that flew off the meteorite in the atmosphere, before it hit ground in the desert, could have survived.

THE PASSION OF NININGER

Harvey H. Nininger (1887–1986) was a professor of biology and field geology at McPherson College, Kansas. In August 1923, at the age of 36, he chanced on an article by Professor A. Miller of the University of Kentucky. In this article on meteorites, published in the popular science periodical *Scientific Monthly*, Miller explained in particular that the probability of observing a meteorite as it fell was less than *one in a million*. On the evening of November 23, 1923, as he was returning home, talking to a friend, Nininger was the privileged witness of the very rare "meteor" phenomenon accompanying the entry of a meteorite into the atmosphere: a cylinder of fire with a diameter comparable to that of the Moon but far brighter than the brightest moonlight. Instantly he decided to search for the "fallen" meteorite from this meteor, but never succeeded in finding it. Yet this decision, the full consequences of which he could not have foreseen at the time, changed his whole life and turned him into the most enthusiastic of meteorite hunters. Resigning his college post, he gave himself up full-time to this new devouring passion and wagered that he could make a living from it.

Beginning with this first meteorite hunt in 1923, he started to develop the technique of communication that he had invented himself and turned it into a highly effective collecting tool. When he arrived at the presumed area of the fall, he would go into a church and, with the permission of the minister, climb into the pulpit and give the congregation a three-minute talk on the details of the meteorite before the Sunday sermon. He asked those present to meet him after the service in the hope that a few eyewitnesses would bring him a fragment of the meteorite or at least some information sufficiently precise to guide his search. He then improved on this technique by using the local newspaper to describe meteors and meteorites and giving talks in schools on the subject. He would not hesitate to go up to a farmer in the street or raise a glass with gold diggers in saloons to tell them what he was chasing so keenly.

The method was successful. On one occasion, for example, he found a meteorite with the aid of a teenager who had been hunting squirrels and was frightened by a hail of stones striking the foliage above his head. After Nininger's visits and talks, it was rare that some farmer, cowhand, or Indian would not bring him a few meteorites during the next few months. On a map of the United States that shows meteorite falls, it is easy to pinpoint the locations where Nininger lived and worked between

1930 and 1950. You only have to look at the areas with the most dots.

Indeed, a single lecture delivered with much enthusiasm immediately converted the listeners into avid meteorite hunters, no doubt because Nininger told them he was ready to pay a handsome bounty for these extraterrestrial objects. Until the discovery of the Antarctic meteorites, he was the greatest meteorite hunter in the field, establishing the record of 218 distinct observed finds and eight falls, adding up to about 2,000 meteorite fragments after thirty years of hunting. To this "direct" collection were added the meteorites he purchased or traded. Thus, his collection included over 700 finds and 150 falls, comprising over 5,500 meteorite fragments. His record of falls has never been surpassed, but his record of finds was beaten by a hunter of Antarctic meteorites, William A. Cassidy, in 1976.

Nininger resigned his position as a professor of biology in 1930 at the worst possible time, in the midst of the Great Depression. He dragged his wife, Addie, who became his most assiduous assistant, and his three children through a difficult, nomadic life always riddled with debt. A year after his resignation he was offered a new and stable university post but, as he would have been required to teach full time, he turned down the offer. To survive this period of his life he came up with an incredibly resourceful and clever way to extricate himself from difficulty and yet pursue his passion. He took a job as a truck driver for a pittance, after securing the permission of his employer, who was also a meteorite fan, to stop en route to give lectures and look for meteorites, which he then had to share with his employer. After vainly trying to make a decent living from his passion by operating a small meteorite museum built near Route 66, which brought visitors near the Barringer Crater, Nininger reluctantly decided to sell his collection, taking the greatest care that it not be scattered in the hands of private collectors. He was then seventy-two.

After lengthy negotiations, the British Museum in London was the first buyer in 1958 (for the sum of about $160,000) of polished sections of 1,200 meteorite samples. Others were then quickly bought up, mainly one by one, by Arizona State University in Tempe, which brought him about $300,000, by the Smithsonian Institution in Washington, D.C., the University of Mainz, and the Natural History Museum of Vienna, Austria. Of the great institutions of this type, the Museum of Natural History in Paris was the only one to hold itself aloof from these highly interesting transactions, due to the disinterest of its successive directors since 1960, who made the mistake of never believing a meteorite collection to be one of their priorities.

Several anecdotes from the fascinating book written by Nininger (*Find a Falling Star*) deserve attention. The first illustrates the intensity of his passion. When he was thirty-four years old, his doctors diagnosed heart trouble that was sufficiently serious to forbid him any vigorous physical activity such as ascending stairs, bicycling, and mountain climbing. He never heeded this advice, caught up as he was in a constant whirl of expeditions, lectures, and fund-raising. Could it be that forgetting about his heart enabled him to live to be nearly a hundred?

The second anecdote illustrates his legendary tenacity. During the Great Depression, he found the necessary strength of conviction to borrow money at a time when he was already deep in debt to buy meteorites at a very good price, which his direct competitors were having trouble selling because of the difficult economic times.

The third illustrates the difficulty of finding meteorites even in as favorable a region as the desert. Having sold off much of their collection, Nininger and his wife took a very active retirement, giving in to another passion: journeys to far-off places. During trips to Indochina and then Australia, Nininger became interested in other mysterious objects that looked like large solidified droplets of very black glass in a great variety of shapes called "tektites," which he collected first in Indochina and later in Australia. During the last voyage, he strode across several deserts on the lookout for tektites, particularly in Nullarbor Plain, which we shall refer to again later. He found neither meteorites nor tektites. Since then, the German George Delisles and the Austrian Christian Koerberl have each found a hundred or so meteorites and nearly a hundred tektites in three weeks of walking in March 1992. Their success is doubtless accounted for by their training in detecting tiny meteorites during field trips to

Antarctica, while Nininger had never been concerned with such meteorites.

WILSON'S PAINSTAKING EXPLORATION OF ARID ROOSEVELT COUNTY

Another collection from a hot desert was assembled by Ivan E. Wilson, who worked for the private meteorite laboratory founded by Nininger (the American Meteorite Laboratory in Denver, Colorado). In 1957, against all prevailing opinion, Wilson decided to search every square yard of a brush-covered desert-like area in the American Southwest. He discovered a total of nearly 160 meteorites! This work, of which meteoriticists remained unaware for a long time, contributed decisively to proving that hot deserts could be treasure troves of small meteorites as long as you walked slowly.

This area, in Roosevelt County, New Mexico, Wilson paced step by step, first collecting seventy-four stony meteorites. In five years, with the aid of two colleagues, he searched through close to 1,000 hectares [2,500 acres] of this desert in the same way, on foot, and collected another eighty-five meteorites. The largest meteorite in this collection weighs about 15 kg [33 lb] and the smallest, about ten grams [1/3 of an ounce].

It is assumed that this large number of meteorites may be attributed to the effect of winds during the catastrophic drought of the 1930s. This period, whose legacy was the Dust Bowl, was the ruin of farmers because the richest layers of topsoil were swept away by the winds, leaving behind hard, bare, light-colored terrain. During this time, fields disappeared in a whirl of dust but meteorites, originally scattered throughout the soil, were uncovered and could be collected on the hard soil that remained. It has recently been determined that meteorites had accumulated over a period of nearly 50,000 years in this area.

It could be that satellite-prospecting techniques will in future allow such zones swept bare of dust by the winds and favorable for accumulation of meteorites in the hot deserts to be identified. The immensity of certain desert areas on the planet still holds the promise of fabulous harvests.

METEORITE SMUGGLING: A LUCRATIVE BUT RISKY TRADE

There has always been meteorite trading, but it has latterly become lucrative. For this reason, smugglers are now taking huge risks to go collecting themselves in certain favorable zones such as the Australian and Sahara deserts. One of the most efficient smugglers succeeded in collecting over 270 meteorites all by himself in the Sahara during two periods of two months each in 1989 and 1990, driving a 4WD vehicle. Thus he beat the record set by Nininger, who spent close to thirty-five years of his life at it, but never found as many meteorites in four months. The efficiency of this smuggler can only increase in the future if collecting expeditions can truly be conducted with two 4WDs.

Certain smugglers have attempted to pull off spectacular individual "heists" of very large meteorites, which are valuable because of their size. One of these schemes occurred in 1990 and was described to me by William A. Cassidy. It took place in Argentina, where one of these smugglers had decided to buy an iron meteorite weighing several tons. This man had rented a crane to load the meteorite onto a truck and succeeded in driving several hundred miles before he was stopped by the police. When he was interrogated, a large sum in dollars was found on him as well as a bill of sale attesting that he had made a $25,000 deal with a certain individual who had no ownership rights to the meteorite. He confessed to the police that he was acting on behalf of a private Japanese collector who had promised to pay him $500,000 for going on this luxury shopping trip. The latest is that he has been released, probably after putting up a sizeable bail. I wonder whether the wily seller of the meteorite has ever been found.

Smugglers risk years of imprisonment because most countries have forbidden meteorites to be exported. But meteorite trafficking is well organized. One European dealer has actually set himself up in business in a Swiss town; we receive a catalog routinely updated with typical Swiss thoroughness. Scientists are compelled to purchase fragments of these smuggled meteorites to prevent them from disappearing into a private collection forever. Hence we are supporting this illicit trade in spite of ourselves.

There is a risk that the breakup of the Soviet Union will intensify the greed of smugglers, and highly valuable meteorite collections will be dispersed. How can we fight in this bidding war conducted under wraps?

The selling price of a meteorite is generally established by its rarity, its state of preservation, and its size. We have seen that one Japanese collector was ready to pay $500,000 for a large meteorite in Argentina weighing several tons. A meteorite picked up after a fall is worth more than a "find" of the same type. The most abundant meteorites (ordinary chondrites), which account for about 85% of the falls, currently go for less than $2 a gram.

But if you find a small meteorite weighing a few grams but of a very rare type such as a type I carbonaceous meteorite, it will fetch over $200 a gram, or more than the price of gold. If your little meteorite belongs to no known group, its price will skyrocket. We saw this with the only lunar meteorite found by a smuggler in Australia, the only such one available for private collectors. Its price is estimated at $1,000 a gram. A famous U.S. smuggler recently managed to acquire a small (some 200 g) Martian meteorite worth "millions of dollars." But you will have to find a private collector because the curators of the major terrestrial collections cannot afford such prices. In general, they trade among themselves. For example, my colleague, Géro Kurat, curator of the Austrian meteorite collection, set himself a maximum price of $100 a gram. He will go no higher, and waits patiently for our "institutional" expeditions to bring back another example of the rare types.

From time to time, when a superb collection such as Nininger's (valued at over $1 million) goes on sale, institutions may receive an exceptional grant from a university or government agency to snap up a lot, provided the heirs of these collections have had them in mind. Otherwise, only the wealthiest private collectors are in the bidding.

Our only effective means of counteracting meteorite smuggling is to organize our own collecting expeditions in hot or cold deserts, with the agreement of the local authorities and the collaboration of scientists in the host countries, who are really far more efficient than the smugglers. We will now describe these modern collecting expeditions, following a path that will take us first to the cold deserts of Antarctica, then the hot desert of the Nullarbor Plain in Australia.

CASSIDY AND THE ANTARCTIC METEORITES

William A. Cassidy had a geologist's training. He began to be interested in extraterrestrial material when he studied the great impact craters on Earth, an interest which took him to Australia to study the Henbury Crater. Another of his exotic destinations was Mauritania, where he met Théodore Monod, who gave him some practical information for exploring an impact crater located near the site where Monod himself had for a very long time (until 1990) been searching in vain for a giant iron meteorite near Chinguetti.

In 1973, Cassidy, then a professor at the Planetary Sciences Department of the University of Pittsburgh, attended a scientific meeting where he heard a paper that would change his entire career. The paper described the accidental discovery in January 1969 of nine Antarctic meteorites by a Japanese geologist participating in a seismic survey campaign near the Japanese base located in the vicinity of the Yamato mountain range in Antarctica. This discovery went unnoticed at the time because the planetologist community was too busy preparing for the return to Earth of the first lunar samples. At this meeting, Cassidy immediately had a comprehensive vision of a spectacular accumulation of meteorites at the surface of certain areas of blue ice where ice flows were slowed down by obstacles. He dreamed of annual collections of several hundred meteorites in such "stagnant" or "dormant" ice zones.

Obviously, the first step was to obtain funding from governmental authorities. So he wrote an initial study grant proposal in 1974, expecting that it would be turned down by the scientific evaluation committee. His fears were justified: the proposal was rejected two years in succession under the pretext that the project was dangerous and the search would be fruitless.* His proposal was finally accepted two years later simply because, contrary to the predictions of the experts, three years after the 1969 exploration another Japanese expedition found 663

* The same two reasons were given by the French authorities when they rejected my first proposal for collecting micrometeorites in Antarctica in 1982.

meteorites in a single month, which certainly showed that Cassidy's proposal was well-founded.

This is another instance of time wasted because of the obscurantism of certain colleagues, or their desire not to share a valuable funding source earmarked for other projects already under way, with themselves probably on the receiving end. New ideas always have a lot of trouble competing with operations that have often become routine but which provide work for laboratories frequently organized and operated today like small businesses. They need a well-filled order book, otherwise bankruptcy stares them in the face and the staff start to complain. Business has its place, but it has nothing to do with the original ideas of small independent teams that have no businesses to support them.

It should be added that the business viewpoint is even more pronounced in France than in the United States, thanks to the actions taken by various people who have headed French research over the last two decades. More serious yet: this research orientation is valued still more highly by the European bureaucrats in Brussels. In a strongly worded speech, former President Bush recently said that along with large projects one should not forget to support small teams because they are often the ones responsible for innovation. As we in France generally lag a few years behind the United States, we might hope that the situation will improve for the young scientists who will be taking our place a few years down the line to prepare for the return to Earth of samples from comets and Mars.

Cassidy's first three expeditions were carried out in collaboration with the Japanese. The meteorites were simply apportioned at random: "One for you and the next one for me." After a single interruption in 1989 due to violent magnetic storms churned up by solar activity, which threw off the navigational instruments of the aircraft carrying the teams and their equipment in the field, the American projects took place every year with four to eight team members, including a New Zealand and/or a U.S. mountain guide (John Scott). The biggest harvest (over 1,100 meteorites), which represents the absolute world record, was gathered in the short space of five weeks in 1990.

The international team meets at Christchurch on the South Island of New Zealand. From there it takes off in U.S.A.F. planes and, after a nine-hour flight, lands at the U.S. base at McMurdo Sound where over 1,500 people, a third of them women, spend the Antarctic summer. After a few days of gathering equipment, food, and fuel, other aircraft take the team to the chosen location from which the hunters will fan out. One of the tactics used on the moving ice far from the mountains and beds of moraine (rocky debris deposited by the ice flow) is to move in a group with each team member on a personal snowmobile. Each snowmobile pulls a sled on which is stowed the equipment and food, selected at the McMurdo station, for each team member.

Aerial photographs are used to pick out zones of moving ice swept clean of surface snow layers by winds, and blue in color, contrasting with the snow. Bereft of their snowy shrouds, the meteorites stand out in black against this light blue background. In such a zone, one generally selects a zone where the very slow speed of the ice flow has slowed down still further because it has encountered an obstacle—the tip of a mountain still buried in the ice, for example. Since evaporation is rapid here, meteorites tend to accumulate in large numbers; they are initially trapped in the ice several hundred miles upstream before they plunge into the depths of the glacial ice cap.

In these zones the snowmobiles are ridden in parallel lines fifteen yards or so apart at a speed of 5 to 10 km an hour [3-6 mph]. As soon as anyone spots a black stone or a stone that looks different from the local rock, everyone stops. The position is noted, the stone is photographed and then placed in a sealed, sterile plastic bag. The day starts at 9:00 A.M. and ends at about 5:00 P.M. with the pitching of tents if the team is camping out, or a return to base camp. To get from one area to another, you may have to do 60 miles a day, so you have to open the throttle to make good time!

But meteorites are also found near the Antarctic mountains, in moraine beds full of terrestrial stones. Here you go on foot, about 5 to 10 kilometers a day.

Then the *frozen* meteorites are taken by plane to a NASA laboratory at the Lyndon Johnson Space Center in Houston, where the lunar samples brought back by the Apollo missions as well as the stratospheric micrometeorites we shall discuss later are stored and preserved for future generations. Once there, the objects are classified after preliminary analyses. The catalog prepared and updated in this way is distributed to all specialists in extraterrestrial material, who are allowed to obtain samples if they send in a proposal which is evaluated by a committee of experts. After spending a few years in Houston, the meteorites are sent to the largest natural history museum in the world (the Smithsonian Institution in Washington, D.C.), whose collection far outranks all the others.

One cannot help being impressed by the booty from the Japanese and American expeditions in Antarctica. In a decade and a half, the number of meteorites in the big international collections has doubled. Today we have nearly 20,000 fragments corresponding to about 3,000 separate falls (meteorites often break up in flight).

In fifteen years of activity, Cassidy became without doubt the greatest meteorite hunter of all time. It is a delight to hear him recount the numerous adventures on his expeditions, which he tried to make truly international, particularly by inviting many foreigners to participate. So, in his way he celebrated the fellowship of humanity well before Jean-Louis Etienne used it as the main audiovisual drawing card of his crossing of the Antarctic.

ALEX BEVAN AND THE NULLARBOR PLAIN IN THE AUSTRALIAN DESERT

The first two Australian meteorites were discovered in the middle of a desert in March 1965 by two geologists who were making topographic surveys. Their attention was drawn to two black shapes which proved to be two fragments of the same iron meteorite, the larger of which is the biggest meteorite ever found in Australia. It can be seen at the Western Australian Museum in Perth under the name of *Mandrabilla*; it weighs 17 tons.

Since 1986, the scientist Alex Bevan from the same museum has systematically organized yearly meteorite collections in southwestern Australia, a very arid desert shunned by the aborigines because there is so little water (a porous dolomite layer drains away the water very rapidly after the few rainfalls). Here, covering an area almost as large as Texas, is the Nullarbor Plain desert—the name is derived from a contraction of two Latin words, *nulla arbor*, meaning "no trees." Collections are made either in 4WD vehicles at a speed of 4 km an hour [2.5 mph] with a distance of 10 m [33 ft] between vehicles, or more slowly on foot, covering about 10 km [6 mi] per day. In 1991 the Australian collection contained about 3,500 meteorites representing 270 individual falls. In 1992, Bevan's team collected another 440 meteorites as part of a collaboration with the European program known as Euromet.

THE EARLY EUROMET EXPEDITIONS

One cannot but be impressed by the quasi-military logistics deployed in Antarctica to collect meteorites, at a cost to the Americans of several hundred thousand dollars per year. Other than a passion for meteorites, one might well ask what drives scientists to keep searching for new ones. Allow me to explain.

The hot deserts are a favorite hunting ground for smuggler "tourists." To counteract their meteorite hunts in this terrain, the Euromet* program supported a meteorite expedition organized by Alex Bevan in March 1992 in the Nullarbor Plain desert. This expedition was highly successful: in three weeks, an international team of eight researchers, five of whom were European, picked up nearly 440 meteorites, representing twenty to forty individual falls. This is a current world record for hot deserts. Two of the team members, who had hunted in Antarctica and whose eyes were particularly well trained, entered into a friendly competition: the first of them, George Delisle, found 110 meteorites and the second, Christian Kœrberl, found 100.

Also, in collaboration with Jean-Pierre Fontes, a sedimentologist who tried to retrace the past climate of the Sahara from soil core samples, we organized an expedition under the auspices of Euromet that was planned for February 1992 but had to be canceled in the end because of an uprising by the Tuareg. We hope the postponement will not last for long, because time is pressing. It would only take a single trip to the desert to assemble a collection of meteorites that has accumulated

* This program was accepted by the European Community in March 1990 because of the energy and competence of Colin T. Pillinger. It combined a group of about forty laboratories and nearly 170 researchers. Its goal is the collection and multidisciplinary study of meteorites and micrometeorites found in hot and cold deserts.

over tens of thousands of years, and remove it for scientific investigation.

These desert collections are priceless. They are part of the legacy of the planet and should be accessible to any competent scientist who might assist local governments in developing research or preservation activities for future generations. Euromet is proposing to train a few researchers in these countries in the most advanced analytical techniques, which could also be applied to many other areas of basic and applied research.

THE HUNT FOR MICROMETEORITES

For over a century, numerous groups have tried to collect micrometeorites on Earth from sites as varied as the garden of an Étretat villa on the Channel coast, the snowy slopes of the Alps, the Antarctic ice, deep sea sediments, lake bottoms, sand beaches, roofs of buildings, the upper atmosphere, and so on, but with no success until 1970. In the most favorable cases, these collections yielded a small number of melted magnetic micrometeorites given the name *cosmic spherules*.

As we explained, we are dealing with the combined difficulties of the extreme rarity of micrometeorites and the enormous pollution by grains of terrestrial dust when they are collected on Earth. Another obstacle is their very short life expectancy due to corrosion in the natural environment; in temperate climates, a stray micrometeorite on a roof will probably not last more than a few months.

For a long time it was thought that the major obstacle to collecting outside the Earth's atmosphere (from a satellite) would be destruction of the micrometeorites on impact when one attempted to pick them up on a collecting plate, since there is no atmosphere to slow them down. But as we shall see, there is a very troublesome source of pollution in space too, which will increase in the future: human trash left behind in space.

Scientists have solved most of these problems with a good dollop of ingenuity. But they needed perseverance as well, for micrometeorites should be entitled to take their place among the big, important meteorites. Indeed, conventional meteoriticists have become accustomed to having sufficient quantities of meteorites available to analyze; in the case of micrometeorites, such analyses are impossible because of their very low mass (one-millionth of a gram). To study them, one has to develop a whole arsenal of micromanipulation and microanalytical techniques, which are difficult, expensive, and require a very lengthy and specialized training before they are mastered.

We will consider only the "modern" collections made since 1970, proceeding first from the middle of the Pacific to Greenland and to Antarctica, then out into the stratosphere, then to 300 km [186 mi] above the Earth, and finally to the tail of Halley's Comet, 150 million km [93 million mi] away.

THE FATHER OF THE "MICROMETEORITICISTS" FIGHTING WITH MARINE SEDIMENTS IN THE MIDDLE OF THE PACIFIC

The first successes of micrometeoriticists may be attributed to the relentless eagerness of a young astronomer, Donald E. Brownlee, who revolutionized this type of meteorite gathering starting in 1970, at a time when scientists were still highly skeptical as to the validity and usefulness of previous collections. Brownlee may be said to be the veritable father of micrometeoriticists. Before him there were no definite criteria for deciding whether or not a given particle was extraterrestrial, and many so-called micrometeorites were in fact black industrial dust (fly ash) which had managed to scatter rapidly over great distances. Brownlee used the latest microanalytical techniques developed to analyze tiny quantities of lunar samples, as well as the most recent results from meteorite studies, to better identify the extraterrestrial origin of micrometeorites. He also "industrialized" these collecting expeditions to gather very large numbers of extraterrestrial particles.

After various balloon experiments which he had planned since 1960, one of which was known as the "vacuum monster" (a sort of giant vacuum cleaner suspended from a balloon flying at an altitude of 20 km [12 mi]), he built a large magnet weighing 300 kg [660 lb], fastened it to a cable over 5 km [3 mi] long, and dragged it across the ocean depths in the middle of the Pacific, 5,000 m [16,500 ft] down. These zones suffer very little contamination from continental dust entrained by winds, because they are thousands of miles off the continents. Once again, success came only after every kind of setback, for example: a cable that kept breaking (the dredging rake got lost and had to be rebuilt) or treacherous currents which gave the mistaken impression that the cable was no longer taut and that the rake had reached the bottom. (At this point they decided to start dredging while the rake was still floating in the water carried by currents, leading to days of phony dredging at a cost of $10,000 a day!) Finally, after a few expeditions that began in 1975, Brownlee and his associates succeeded in a few years in "magnetically" raking up over 100,000 partially or completely melted micrometeorites. Their highly magnetic character is accounted for by their fusion followed by oxidation in the atmosphere, producing magnetite crystals.

GIGANTIC, INEXHAUSTIBLE MICROMETEORITE COLLECTORS IN GREENLAND AND ANTARCTICA

Ten years later, working with two glaciologists, the Dane Claus Hammer (Geophysics Institute of the University of Copenhagen) and the Frenchman Michel Pourchet (Glaciology Laboratory of the CNRS,* Grenoble), my team and I were able to renew the collecting of micrometeorites on Earth, thanks to the development of new methods for working the great glacial ice caps of Greenland

* French National Scientific Research Center.

and Antarctica as they were, giant, ultraclean, and *inexhaustible* collectors of micrometeorites.

Our first proposal was submitted without success to the French government in 1982; it concerned a micrometeorite expedition to Antarctica. Our second proposal, submitted in 1983 for an expedition to Greenland, had a vigorous proponent in the person of a clear-sighted radioastronomer (Jean-Louis Steinberg), but the French meteoriticists merely smiled ironically when I presented my proposal to the scientific selection committee.

Our project actually materialized in July 1984. After two years of wrestling with the selection committees, I suddenly found myself with four Danes and appropriate camping gear, all dropped from a helicopter, 20 km [12 mi] from the margin of the Søndre Strømfjord glacier on the west coast of Greenland in what is called the "melt zone" of the ice sheet. I was petrified by the dazzling spectacle, by the fear of Nature at its most hostile, and by the hope that it could be our most valuable collaborator.

Greenland explorers have always avoided crossing this zone during the very short Arctic summer, which lasts from mid-June to early August. During this time, the ice melts, releasing an enormous quantity of water that streams down in torrents and fast rivers that may reach speeds of 3 m per second [10 ft/sec]: a dangerous labyrinth from which it might be difficult to escape. There was only one solution: a helicopter, provided one could find an experienced pilot who would agree to take a few risks. Such pilots are few and far between. We camped out for ten days on ice that had been riddled into a Swiss cheese by innumerable holes 20 cm [8 in.] deep, of all diameters, into which water trickled ceaselessly, revealing bottoms coated with a very fine layer of black mud (called "cryoconite").

We had made a rather crazy bet. We knew that the torrents poured into small lakes at the bottom of the ice valleys which formed and reformed every year at the same spots, because they reflected the hill-and-valley relief of the bedrock over which the glacier flowed. This bed was under a layer of ice about 1 km [0.6 mi] thick. About 1 m [3 ft] of ice melts per year. If the collecting basin is several miles across, hundreds of millions of tons of water would pass across the lakes every year, depositing dust initially trapped in the ice. This process could repeat for hundreds of years, representing over a billion tons of water. Our deduction was that actual deposits of sediments rich in micrometeorites should be found on the bottoms of these lakes.

The helicopter had to fly around for some ten minutes to find a landing spot that was not too dangerous, and we became lost. Specifically, we could not find any trace of a superb "blue" lake which we had carefully pinpointed on an aerial reconnaissance photograph. It took us several days of walking to discover, to our great relief, another blue lake with splendid lodes of black sediments. Henrik H. Thomsen and I, dressed in trout fishing suits, roped ourselves together, and walked out onto the slippery lake bottom in a swift current. Using a hose connected to a plastic hand pump weighing about a pound (usually used to empty the heads in pleasure boats), activated on the shore by our colleagues, we vacuumed up the black sediment from the lake bottom.

In this way we collected about 20 kg [44 lb] of cryoconite in two days, which contained, as we discovered later in the laboratory, numerous micrometeorites, particularly unmelted ones which are very rare in collections extracted from marine sediments. These micrometeorites were also a thousand times the size of those collected from the stratosphere. Our commando-style operation had succeeded, and for Claus Hammer and myself it was the finest memory of all our expeditions; moreover, it was my entry into the world of micrometeoriticists, and hence the starting point for this book.

There were other cryoconite-collecting expeditions, particularly in July and August 1987, once more with the Danes. They were part of a far more ambitious project with thirty participants and two base camps, entitled: "Ten Thousand Years of Micrometeorite Falls in Greenland: A Glaciology, Climatology, Volcanology, and Planetology Project on the West Greenland Ice Cap."

However, the most fruitful Greenland expedition was not one of those I organized with great difficulty with Hammer, but rather the extraordinary adventure of a group of twelve-year-old boy scouts who met up through the pairing of a French village with a Greenland village. The scouts used to meet every year during the summer vacation, one year in France and the next in Greenland. In 1988 it was the turn of the French boys to go to Greenland and meet their fellow scouts for a field trip led by their guide, André Drouin. The project was to repair a hut that the French polar expeditions had built in 1958, when they found a very safe, crevasse-free access route near de Kervains harbor over which they could drive their tractors on the Greenland ice cap and make the crossing.

I met Drouin and talked him into bringing a group of scouts at least 5 km [3 mi] inside the glacier, avoiding the crevasses. Drouin was a good guide and they went even further, 9 km [5.5 mi]. Enjoying themselves hugely, they brought me back 2 kg [4.4 lb] of cryoconite. As luck would have it, it was found to be far richer in micrometeorites than the samples we ourselves had collected first in 1984 and then in 1987, after long and painstaking preparation. I remember I had to make the rounds of several sedimentology laboratories to borrow equipment for raking the bottom of a lake. There was one wolf-trap affair which you had to throw into the water on the end of a rope, but first you had to pry apart its wicked-looking jaws without cutting your fingers. The idea was that as soon as the trap hit bottom the jaws were supposed to close and trap sediment. But they always snapped shut before hitting the water. As I said, when you're hunting micrometeorites you can't be too fussy and you have to think on your feet: we ended up using the head pump from a boat, quite successfully.

Nineteen eighty-seven was a very busy year. In July and August, for six weeks, we went out on a new expedition to Greenland with our Danish colleagues. In December we set off for Antarctica with two French glaciologists from the Glaciology Laboratory of the CNRS in Grenoble to melt down nearly a hundred tons of blue ice in the field* at Cap Prudhomme, about 6 km [4 mi], as the crow flies, from the French station at Dumont-d'Urville.

To conduct this operation in the field, we used a sort of rudimentary micrometeorite "factory" developed by the glaciologist Michel Pourchet. The factory was composed of two oil-fired steam boilers set to produce jets of hot water at 70°C [158°F] and used to melt the ice. So at the end of the day you get two pockets of water with a few cubic meters [several hundred gallons] in each. You draw up this meltwater with a pump and filter it through steel sieves, which gives you a sort of very fine glacial sand.

This sand was far richer in micrometeorites than Greenland cryoconite. In particular, there were about 100 mg of a sand with a grain size between one twentieth and one tenth of a millimeter and, to our great delight, *one grain out of ten was an unmelted micrometeorite*. An absolute record of purity for micrometeorites collected

* In Antarctica, during the warmest months of the southern summer (December and January), it is much colder than in Greenland during the Arctic summer (June and July). Except for a few days a year in January, when the temperature goes a few degrees above freezing for a few hours a day, the ice stays frozen even near the shore. So to get water you have to melt the ice with all kinds of contraptions.

on Earth! This purity allowed us to find magnetic particles and nonmagnetic particles and discover a very interesting population of micrometeorites, very friable and very primitive, practically absent from the Greenland collection. In 1991 we returned to Cap Prudhomme to melt another 260 tons of ice with three steam boilers and to try out some new equipment. The result was another 200 mg of this very fine sand which contained a total of several tens of thousands of micrometeorites over fifty microns (μm) in diameter.

HIGHER AND HIGHER STILL

Let us now leave the surface of the Earth and reascend into the atmosphere. After his first magnetic raking of marine sediments in 1975, Donald Brownlee, in collaboration with a NASA team, made a very different collection of samples the following year, a project which is still going on today because it proved of great interest. A small 10-cm^2 [1.5-$in.^2$] plate was mounted under the wing of a U2 spy plane. This plate was covered with a layer of extremely viscous silicone grease. When the aircraft reached cruising altitude, about 20 km [12 mi] up, the plate was exposed. In a few hours of flight, at a speed of 900 km an hour [550 mph], it swept an enormous volume of air. Most of the tiny grains (about 10 μm in size) floating in the air stuck to the silicone grease. Back on Earth, the particles were extracted one by one from this grease, which was rinsed off in liquid Freon.

Until 1989, numerous flights were made yielding a collection of about a thousand very fine and highly interesting micrometeorites called "interplanetary dust particles." The project was interrupted only during major volcanic eruptions such as that of Mt. Pinatubo in 1991, which polluted the stratosphere with too much volcanic ash, and no interplanetary dust particles could be found on the U2 plates. A collector ten times the size of Brownlee's first one made its first flights successfully in 1990. In the near future this should give us "large" micrometeorites with diameters of fifty µm which in the past have been very rare (in fact, only one such particle had been found by 1990) in this stratospheric collection. The 200 mg of Antarctic sand collected in five weeks in 1991 already contained several tens of thousands.

But this is not all. "Space" meteoriticists go even higher to collect micrometeorites, or more precisely to collect their impact craters made in plates of different kinds, called "impacters," exposed at an altitude of about 300 km [185 mi] before the atmosphere has slowed the particles down. The particles slam into the plate just like those that bombard the Moon and leave behind microcraters as they self-destruct. In certain favorable cases, some residues of the incident particle remain in the bottom of the crater and they can be analyzed by highly sophisticated techniques (analytical microscope).

The first detailed study of this type was conducted on the thermal (aluminum) protection shutters of the American SMM (Solar Maximum Mission) satellite, which was launched to study solar activity and was recovered by astronauts in space in 1984. The last study was performed as part of the LDEF (Long Duration Exposure Facility) space experiment during which plates were to be exposed for just under a year in space, but which remained in orbit for over five years because of the explosion in flight of the space shuttle *Challenger*. The most promising LDEF impacters, currently being studied, are highly pure germanium plates about 2 m^2 [21 ft^2] in area.

One of the unexpected difficulties of these experiments is that space 300 km up is highly contaminated with pieces of trash orbiting around the Earth which themselves produce craters in the impacters. But this space pollution, which can only get worse in the future, is essentially dominated for the time being by tiny particles less than a few thousandths of a millimeter in diameter. So we are still hopeful of gaining useful information on the larger micrometeorites. But since these particles are very rare, to capture just a few of them large collectors exposed for at least a year to the micrometeorite flux are needed.

In March 1986, the meteoriticists went still further from Earth on a day when a cortege of six space probes made their historic rendezvous with Halley's Comet 150 million km [93 million mi] from Earth. The Giotto and Vega probes were each equipped with a "time-of-flight mass spectrometer." This determined the approximate chemical composition of the microscopic jet of gas produced by the grains of dust expelled from the comet when they impact a metal plate at about 60 km per second [37 mi/sec] and volatilize. Thus, for the first time, we have a semi-quantitative analysis of true cometary micrometeorites. One of the surprises from these results was the discovery that the cometary micrometeorites were markedly enriched with organic matter, which was immediately given the name of CHON because it contains carbon, hydrogen, oxygen, and nitrogen atoms whose respective chemical symbols are C, H, O, and N.

In the future, it is planned to equip the impacters placed in orbit around the Earth with retarders made of a kind of silica aerogel. This is a very light type of gel filled with air bubbles. It would be able to brake the slower particles to permit what the most optimistic scientists are already calling their "intact capture" (meaning that only partially destroyed particles could be picked up). In five or six years, NASA plans to assemble in space the Freedom space station from which astronomers could deploy a large detector, 10 m^2 [108 ft^2] in area, which would determine the trajectories of the micrometeorites just before their destructive impact with the impacters. A chemical analysis of the tiny gas jet produced on impact would be performed by devices probably derived from those on the Giotto and Vega probes. Such a "telescope" would open the field to a new and exciting astronomy, that of dust grains.

STILL MORE METEORITES? WHY?

One may classify the best world meteorite collections by the total number of separate meteorite falls that they contain. At the top of the classification is the Smithsonian Institution in Washington, D.C., followed by the Natural History Museum in London, the Muséum National d'Histoire Naturelle in Paris, and the Naturhistorisches Museum in Vienna. These four institutions have over 20,000 meteorite fragments, the product of about 3,000 *individual* falls. For micrometeorites, the figures are more impressive. For example, in the 200 mg of sand extracted during five weeks of field work from 260 tons of ice at Cap Prudhomme in 1991, which represents less than a quarter of a cubic centimeter of material, we found several tens of thousands of micrometeorites measuring 50 to 100 μm. During the Antarctic summer of 1994, we will return to Cap Prudhomme to double this number. During the Arctic summer of 1995, we will probably meet Claus Hammer again in frozen northern Greenland to try to collect "minimeteorites." Why continue this race for microscopic or massive extraterrestrial materials year after year?

Some of our motivations are of a "technical" nature; they are those that we state as a priority on the proposal forms we submit to the various selection committees to whom we have to turn with increasing frequency to fund our research. These committees, which are multiplying faster than the available funds, are to be found everywhere in our path, like perilous obstacles to be overcome. They are anchored both in the principal French research institutions identified by initialisms such as CNRS, INRA, CNES, INSU, ORSTOM, IFREMER, IFRTP, IN_2P_3, BRGM, CNET,* and in our innumerable government agencies. After ten years or so, they were seen also to multiply in the departments of the European Economic Community in Brussels, where in addition to the rest of the red tape, we have to fight a quota battle for distribution of funding between the various member countries or the various scientific and technical programs according to priorities which are far from clear and which can be called into question by the best-organized pressure groups.

These proposals also suffer from what I call the "ISOCOM syndrome," if I may take my turn in coming up with an acronym, short for "inflation of sophistication in complication." In a recent set of forms from one specific program, here is what I was asked to describe: on Form 1, the precise objectives of my research; on Form 2,

* CNRS: National Scientific Research Center; INRA (National Institute of Agronomic Research); CNES: National Space Research Center; INSU: National Institute of the Sciences of the Universe; ORSTOM: French Scientific Research Institute for Development in Cooperation, formerly Overseas Territories Scientific Research Office; IFREMER: French Institute of Research and Exploitation of the Sea; IFRTP: French Institute of Polar Research and Techniques; IN_2P_3: National Institute of Particle Physics and Nuclear Physics; BRGM: Geological and Mining Research Bureau; CNET: National Telecommunications Research Center.

the research schedule, which comes down to predicting exactly the date on which you will make your discoveries; on Form 3, the methodology I plan to use; on Form 4, the spin-off of the research for society (so it is desirable to provide a list of patents); and on Form 5, a detailed list of international collaborations, with the particular recommendation of collaborating with southern countries (Algeria, Portugal, Spain, Greece, etc.) whose scientific and technological development is to be aided.

In any case, watch out if you're not a member of a well-organized gang which will close ranks so as not to share the loot with a newcomer, if your proposal is not part of a "hot" program launched by some government department (pretty much guaranteed to lack originality), and especially if it appears to be too innovative. In the latter case, even your own colleagues, who will be your judges, will, with a few rare exceptions, fail to understand its usefulness and interest.

Yet real research, the research that excites us, involves diving into the unknown, in frequent encounters with the unforseen, to satisfy curiosity but also to find truth, when one doubts a scientific dogma. As soon as we have understood the problem to the point where we can fill out the proposal forms properly, the research is actually completed! We have to continue our march into the unknown, where we are so inefficient. How can you describe the methodology to be used? To make new discoveries, you will probably have to use a different instrument. How can you give a list of patents when you know that you will certainly wander down some unexplored path? How can you know in advance who will agree to accompany you down this path? Our usual way of getting through this maze is to submit a research proposal along lines we have already partially explored and from which we have a few preliminary results. With a fraction of the funds allocated, we take off for the unknown, always trying to stay ahead of the game. My hair still stands on end when I think about those forms.

Let us return to the technical justifications, which are easy to describe for micrometeorites. We want to improve on the grains we collect so that they contain more and more micrometeorites and fewer and fewer terrestrial grains. But it is impossible to predict exactly where the best deposits are located. You have to prospect in the field, in ice that is far cleaner than the usual laboratory environment, to test the material under extreme conditions, to identify the least contaminated zones, and to learn how to reduce the contamination caused by our own activities. What is more, each collection of micrometeorites that we bring back—whether from the bottoms of oceans, from ice, from the stratosphere, or from space—has its own advantages but also its own limitations. It is by comparing these different collections that we can hope to define these limits and discover the "unvarying" characteristics of micrometeorites that depend neither on the site nor on the method of collection.

For example, you have to disaggregate Greenland cryoconite vigorously with a stiff brush to release the mineral grains firmly encapsulated in the cocoons of filamentary bacteria of which cryoconite is mainly composed. So you run the risk of destroying the most friable micrometeorites, which are probably those of greatest interest. Hence the idea of melting a hundred tons of ice on the ice cap using steam boilers. By pumping the meltwater and filtering it through sieves, we should be able to collect the grains trapped in far cleaner ice than that processed in Greenland, and far more gently without fragmenting them further. Moreover, cryoconite cannot develop in Antarctica because the cold is far more intense than in the Arctic. All these favorable factors should contribute considerably to improving the quality of our collections. Once more this guess was confirmed during a new expedition in 1991 with an improved micrometeorite factory, allowing the melting of another 260 tons of ice.

To these technical reasons we must add a more scientific motivation generally accepted by the members of the selection committees who evaluate our "forms": we want to discover rare, new, unknown objects which do not fit into our classifications. For example, we dream of finding objects even more primitive than the most primitive meteorites, which will suddenly open a new window on the early history of the solar system. At the other extreme, we are also interested in finding far more differentiated and younger objects that have lost their primitiveness, and come from planets and their satellites. The interesting thing about such objects is that they would contribute to the tough goal of "comparative" planetology (as yet a remote one), which is to compare the planets and their satellites in order to better understand the formation and evolution of the Earth. To increase our chances of finding these rare objects, we have to multiply the number of meteorites and micrometeorites collected, and solve the thorny problem of rapidly characterizing such a large number of objects.

In one television broadcast, I was told that the cost price of Antarctic micrometeorites, in terms of one gram, was greater than that of a diamond because of their very low mass, representing a total of about 50 mg. This brutally frank remark caught me by surprise and left me stammering. If you pursue this line of reasoning you come to the following conclusions (I will confine myself to three examples): (a) the cost of a thousand very small stratospheric micrometeorites collected by NASA (total mass about 0.01 mg) in a ten-year program using two stratospheric planes and their crews was at least 100,000 times that of the Antarctic meteorites. And yet NASA is continuing with this project! (b) For the MEGASETI program, which is searching for extraterrestrial intelligence, the American taxpayer has to shell out at least $100 million over ten years, and may discover nothing at all. But the project must be continued, because it is one of the most exciting human explorations of our decade. With the best scenario, humanoids will exchange with other "planetoids" a few "packets" of insubstantial radio waves whose weight is "zero." So the price of these packets of waves becomes infinitely high! (c) And what can we say about ultimate building blocks of matter like quarks, intermediate bosons, and other fundamental particles in the atomic nucleus that are synthesized in the huge accelerators of the CERN (European Nuclear Research Center) in Geneva, whose annual funding is about $500 million? Their mass is infinitely small. The price per gram would be billions of billions of billions times higher than that of the Antarctic meteorites.

This is why it has long been accepted that the interest of scientific "material" is not a function of its weight; it would be preposterous to weigh it out like the products we use daily. There is no valuable information in an ingot of gold nor in the largest of earthly diamonds. They are of no interest to science.

By hunting meteorites in Antarctica and the hot deserts, meteoriticists have discovered eleven lunar meteorites in Antarctica.* Theoreticians had predicted that it would be "impossible" to capture lunar meteorites, i.e., fragments of lunar rocks ejected during the formation of impact craters on the Moon. It seemed to them even more impossible to capture Martian meteorites because Mars is far more distant from the Earth than is our satellite.

Before the discovery of the lunar meteorites, a few daring meteoriticists believed that half a dozen meteorites of recent age could have come from Mars. But since

* We still do not really understand why lunar meteorites (chunks of the Moon) are only found in the Southern Hemisphere. A twelfth lunar meteorite was discovered in an Australian desert by a smuggler.

the theorists had decided that there could not be any on Earth, this hypothesis seemed rather precarious. After the discovery of the lunar meteorites, they had to think again, and they are now able to obtain Martian meteorites by "computations," during impact cratering events on Mars. So most meteoriticists are convinced that there are Martian meteorites in terrestrial collections. By studying them, we could do a better job of developing the automatic analytical instruments that will be set down on the surface of Mars in the 21st century. And who knows? Maybe one day we will find meteorites from Mercury!

From this viewpoint, if we are to find new objects from the solar system, it seems important to devote some effort to collecting micrometeorites from the stratosphere, the polar ice caps, and space. Indeed, because of the small size of micrometeorites, their orbits are perturbed not only by the gravitational forces affecting meteorites but also by nongravitational forces connected, for example, with illumination of grains by different types of solar radiation. With the aid of these forces, which would permit wider "cosmic combing," one might hope to put a greater variety of bodies from the solar system into capture orbits by Earth.

Moreover, since micrometeorites slow down higher up in the atmosphere than do meteorites, in a region where the air is very thin, the shock they undergo is far less intense. As a result, very friable micrometeorites may survive while meteorites made of the same material would be destroyed. So the stream of micrometeorites might contain new highly friable objects which would be poorly represented in or absent from meteorite collections; this might be the case with cometary grains considered by most meteoriticists to be very primitive.

This discovery of friable grains was actually made back in 1976 when Brownlee found the interplanetary stratospheric particles; recently we extended the discovery to Antarctic micrometeorites which are far larger. In this collection, the proportion of *primitive, friable* objects, which represent only about 4% in meteorite collections, jumps to nearly 80%. Less than 20% of micrometeorites are hard "rock" samples, like those found in the major families of meteorites.

In this breathtaking, continuous search for new substances, the main motivation of all of us is the thirst for the unknown, our frantic desire to discover unexpected scientific and technical horizons, and new natural beauties. Indeed we share this thirst with all researchers, whether they are looking at minerals, insects, microscopic fungi, flowers, archeological remains, black holes in space, new interstellar molecules, fragments of dinosaur bones, pollen grains fossilized in mud, or new "computer beauties" like the generation of fractals which take shape in vibrant colors on a computer screen. This quest for knowledge is what spurs us on—unless we want to verify a piece of dogma that we are questioning by ourselves, or we are engaged in manufacture (of a semiconductor or superconductor), or designing a new analytical instrument just to make progress without really realizing its limitations or possibilities.*

In brief, to quote the friendly observation of a fellow researcher who became a psychiatrist, we are incurable addicts of the unknown. This is confirmed by the problems we find in our family lives. As long as the technocrats who evaluate our proposals have not personally succumbed to the same addiction, there will be no hope of our receiving simplified grant proposal forms.

For a request for meteorites and micrometeorites collected by the European Euromet program, the necessary application forms were designed by the "addicts" themselves. Result: they are very simple: our requirement is that proposals be in essay form, a *maximum* of one page long, leaving the candidate's talent free reign to write the page.

In the following pages of this book, we will see how meteoriticists succeeded in deciphering very ancient and exciting fossil messages recorded in meteorites, which may be seen as unique, fabulous archivists of our "presolar" and then "nebular" past.

* The inventors of the semiconductor, which revolutionized modern life by enabling developments in computers and home entertainment, were so far from foreseeing its applications when they made their discovery that they did not apply for a single patent.

III

APPROACHING "PRIMITIVENESS"

THE USEFULNESS OF A CLASSIFICATION

From fragments of Roman amphoras, archaeologists are able to reconstitute patterns of trade, manufacturing techniques, and the advance of the Roman legions, provided these fragments are properly classified. Similarly, classification is the first and essential stage in the work of meteoriticists. We will not go into endless technical details but will state the principle, which is to establish whether these families are derived from one common ancestor, the most primitive of all, then place these families in a number of different "cosmic furnaces," even asking what types of meteorites formed the building blocks of Earth, and how these blocks and bricks were baked. This classification of meteorites results from the work of several generations of scientists.

First it was thought that meteorites all derived from one parent body which broke up upon impact. This body had a stratified (layered) structure, like the Earth. One classification is derived from this hypothesis: the core, composed of an iron-nickel alloy, produced the iron meteorites (the "irons"), then the next layer up (the mantle) was the source of the "mixed" meteorites (lithosiderites or "stony irons"), made of a block of iron in which stony inclusions are embedded. Finally, the crust supplied the stony meteorites, or "stones," divided into chondrites and achondrites depending on whether or not they contained chondrules—round objects about 0.2 mm to a few millimeters [up to about 1/4 of an inch] in size, not found in the rocks that make up the Earth. However, this classification was unable to withstand improvements in analytical techniques.

One of the most powerful modern methods, available only since 1973, consists in measuring the isotopic composition of oxygen in stony meteorites, and also in the silicate inclusions of the irons (we will henceforth include the stony irons in this group). Meteorites can then be grouped into families (seven families for the stones and twelve for the irons), which cannot be derived from one another by known processes such as melting. These differences in isotopic composition are accompanied by variations in chemical composition like those of volatile compounds and elements (water, rare gases, zinc, bromine, and others), and rare or trace elements such as germanium and gallium; these variations are particularly useful for subdividing the main groups into subgroups.

Moreover, mineralogists such as Mireille Christophe Michel-Lévy, Géro Kurat, and Martin Prinz characterized the whole mixture of minerals that constitute a meteorite, using not only grain texture observations (size, external shape, porosity, and the like) but also very powerful methods of crystal "chemistry" based on a precise analysis of the individual minerals. These analyses first revealed that about 85% of meteorites contain chondrules. Mere observation of the sharpness of the edge of a chondrule embedded in a matrix of finer grains is enough to give an idea of the intensity of the "thermal metamorphism" they underwent when their parent bodies were heated for a long time (over a million years) at temperatures of approximately 500–600°C [950–1100°F], far lower than their melting points. During this metamorphism, they did not melt; under these conditions, however, the grains of a meteorite can be equated with droplets of highly viscous liquid all packed together which slowly exchange their atoms by "diffusion." Then the sharp edge of the chondrule becomes gradually blurred.

This diffusion simultaneously causes other remarkable effects: for example, a loss of volatile elements or a substantial change in the initial chemical composition of the minerals. Thus we discover that, in the most primitive meteorites, the main silicate minerals (olivines and pyroxenes) are formed under different physical and chemical conditions before they are trapped in the parent body of the meteorite. Their compositions differ markedly from one crystal to another and the assembly of minerals is known as "unequilibrated." On the other hand, in the achondrites, which in fact are lavas, the minerals were formed when a given volume of molten magma cooled at the same place and at the same time. All olivines thus have the same composition and there is a strict relationship between the compositions of olivines and pyroxenes. In this case, the assembly of minerals has become "equilibrated."

Mineralogists have constructed a metamorphic scale that ranges from 1 to 7, which completes the chemical and isotopic classification. In meteorites at 7 on the scale, the chondrules have disappeared and all the olivines have strictly the "same" composition (but these meteorites are not achondrites). On the other hand, meteorites classified 1 to 3 have highly unequilibrated assemblages of minerals and chondrules with very sharp edges.

When we combine all these analyses—chemical, isotopic, and mineralogical—we arrive at a large number of meteorite groups and subgroups. In all, there are nearly twenty distinct groups of stony meteorites and probably over fifty groups of irons. Most of these groups cannot be derived from each other by simple processes like melting or diffusion. Thus they must correspond to several dozen different parent bodies (some authors list about eighty). In this book we are interested only in the most primitive stony meteorites and will set the others aside, interesting as they may be.

The groups of meteorites are divided as follows, according to a decreasing order of "primitiveness": (1) *only 4% of meteorites* simultaneously have unequilibrated mineralogical compositions, a high content of volatile compounds such as carbon (hence the name carbonaceous meteorites), and chondrules with very sharp edges. They constitute the most primitive* group of meteorites. We also find refractory inclusions (made of aluminum, calcium, and titanium oxides), discovered for the first time by Mireille Christophe Michel-Lévy. We can distinguish several subgroups: the CIs, CMs, CVs, COs, and CKs. The CIs are the richest in carbon and contain neither chondrules nor grains of metal. (2) *About 80% of meteorites* are ordinary chondrites and their groups are designated by the letters LL, L, H, and E. The edges of the chondrules are, in general, less sharp and the assembly of their minerals is equilibrated starting with 4 on the metamorphic scale. They also contain numerous metal inclusions. (3) *About 8% of meteorites* form the group of achondrites; this group has acquired a certain notoriety because it contains in particular twelve lunar meteorites and 8 meteorites whose Martian origin is "almost" proved. (4) Finally, *about 6% of meteorites* are "irons" and about 2% are stony irons.

In the micrometeorite collection, we do not find such abundances. In a batch of 400 or so randomly selected micrometeorites, over 80% belong only to the primitive meteorite group (representing only about 4% of meteorite falls), and the iron micrometeorites represent less than 1% of this population. Moreover, these rare iron micrometeorites are made of strange objects, much softer than a fragment of an iron meteorite.

* The word "primitive" does not mean "older." It simply identifies meteorites whose various components were the least metamorphosed when they were formed in the solar nebula.

What does this marked difference between the populations of micrometeorites and meteorites mean? It is assumed that the flux of meteorites and micrometeorites in space is dominated by a more primitive material than that of ordinary chondrites, which represent about 80% of meteorite falls. This hypothesis agrees with the preponderance (65%) of asteroids similar to primitive meteorites in the asteroid belt, from which most meteorites come. The rarity of primitive objects in conventional meteorite collections is probably not representative of their true proportion: their greater friability reduces their survival when they are slowed down in the denser layers of the atmosphere (30-80 km or 19-50 mi above the Earth) where a destructive shock wave develops. On the other hand, the primitive micrometeorites, equally friable, are slowed down at higher altitudes (about 100 km [62 mi] up) in layers of very thin air where these destructive shock waves are unable to develop. So micrometeorites would be a more representative sample than meteorites of the flux of extraterrestrial objects traveling in interplanetary space near the Earth.

DESPERATELY SEEKING PARENT

Where do meteorites and micrometeorites come from? We have said that asteroids and comets are possible parent bodies; we also mentioned lunar and probably Martian meteorites, which are chunks of the lunar and

Martian surface, respectively, ejected in space on strong impact.

The lunar meteorites (totalling a dozen) were identified because astronauts have been to the Moon, so these objects can be compared with the rocks brought back by Apollo and their similarity demonstrated. Martian meteorites, just as rare because we have only 8 of them —shergottites, nakhlites, and chassignites—present difficulties of identification. True, we can refer to only a few analyses made by one of the two U.S. Viking space probes, which made a soft landing on Mars on July 20, 1976. Two of these meteorites have a rare-gas composition similar to that measured for the Martian atmosphere by the Viking probe. To this we must add the similarity in iron content of the Martian soil and these meteorites; finally, the extreme youth of these "Martian" meteorites (about a billion years) makes their asteroidal origin highly improbable.

It became possible to compare the chemical composition of major elements of micrometeorites with that of the very fine dust ejected by Halley's Comet when it came close to Earth in March 1986. The chemical composition of this dust was remote-analyzed by several mass spectrometers aboard the Vega and Giotto space probes. Although these were highly qualitative analyses, they strongly suggest that the group of tiny "anhydrous" micrometeorites (with no hydrated minerals like the clays), collected in the stratosphere, have similarities with comet material.

Yet, although the vast majority of meteorites probably come from asteroids, no probe has yet landed on one. The only chemical and mineralogical data come from remote analyses made by telescopes of the spectra of the light reflected by the asteroids. Asteroids are distributed in a broad belt between Mars and Jupiter, the width of which is about four times the distance between the Earth and the Sun. As their composition varies according to the distance from the Sun, we come across a contradiction: the farthest away—about half of them—are dark-colored and related to the carbonaceous meteorites, the most primitive. But these carbonaceous meteorites are far rarer in terrestrial collections (about 4%). The closest, the "S" ("stony") asteroids, are far lighter in color and thought to be made of stone. But none of the other stony meteorites, less primitive, such as the ordinary chondrites which represent about 80% of meteorite falls, strictly resembles them.* So, where are the parent bodies of the ordinary chondrites to be found?

Most meteoriticists, aware of the difficulty, hold that conventional meteorite collections are an incomplete sampling of the small bodies of the solar system. We believe that micrometeorites could be a more representative sample since our most recent determinations indicate that over 80% of them are related to the carbo-

* Another hypothesis suggests that the optical properties of S type asteroids are masked by a superficial layer of dust covering their surface; according to this hypothesis, the dust has undergone various degrees of alteration and prevents the true nature of the asteroid from being identified.

naceous meteorites, which are underrepresented in the Earth's meteorite collections.

Meteoriticists have tried to use other methods to define the nature of the meteorite parent bodies.

We can measure the rate of cooling (after heating) of meteorite parent bodies in different ways. This rate of cooling is an indirect measurement of the size of the parent body: the larger the body, the slower its rate of cooling, as with lava layers on Earth. One of the most promising methods, now under development, is based on extremely precise measurements of meteorite age made by the Gérard Manhès team, with an approximation of one million years. The faster the rate of cooling, the earlier the meteorite formed.

Thus we obtain cooling rates of between about 1°C and about 100°C [2–180°F] per million years. Through appropriate modeling, we can show that such rates do in fact correspond to asteroids less than 100 km [62 mi] across. It is improbable that cometary nuclei could have been heated to sufficiently high temperatures to affect age measurements and yield measurable cooling rates.

Another and totally different method is based on the detection and measurement of the trajectories of meteors and shooting stars. Meteoriticists hoped for a great deal from three camera networks, one of which was built in Canada, one in the Great Plains of the United States, and one in Germany (near the Czech border) to photograph such trajectories. There were two objectives: firstly, determining the complete orbit of the body, which exceeded 1 billion km [620 million mi], by studying the segment of trajectory a few tens of kilometers long (only visible in the atmosphere in the form of a shooting star), producing a tiny straight-line track of a few centimeters in the photographic emulsion; secondly, determining with sufficient precision the point at which meteorites fall to Earth so that they can be collected as quickly as possible. The latter objective proved to be highly utopian because of the difficulty of evaluating the altitude at which the star was "extinguished." When a shooting star looks as if it is falling just behind your back, you sometimes have to explore a zone of 10,000 km^2 [3,850 mi^2] step by step before you find its fragments. After ten years of using these camera networks, scientists succeeded in photographing and recovering only three meteorites (at Příbram, Czechoslovakia; Lost City, Nevada; and Innisfree, Alberta), the trajectories of which indicate very clearly that they *come from the asteroid belt.* We can understand why Nininger did not find his first meteor on November 23, 1923!

However, these networks were not totally useless because they allowed us to define numerous trajectories described by large and small meteorites (not collected). The results are simple: over 90% of micrometeorites measuring over a few millimeters* have orbits like those

* Micrometeorite trajectories are visible provided the micrometeorites are more than a few millimeters in diameter. For the smaller micrometeorites, down to one-tenth of a millimeter, these trajectories can only be observed by radar techniques. For these "invisible" micrometeorites, we always obtain over 90% cometary trajectories.

of comets, inclined to the plane of the Earth's orbit and resembling highly elongated ellipses. On the other hand, those of conventional meteorites weighing over 100 g [3.5 oz] are similar to those of the asteroids that cross the orbits of Mars or the Earth, which are classified in the Apollo, Amor, and Aten families. These orbits, far more circular, are essentially confined to the plane of the Earth's orbit. In this case it is said that they are slightly inclined to this plane, and exhibit small "eccentricities."

Some ten years ago, the situation appeared to be clear. Micrometeorites were considered to be essentially samples of cometary dust, while meteorites were held to be asteroid fragments. This was too simple to last, and at least five problems cropped up: (1) The calculations of Jack Wisdom showed that, under certain circumstances, asteroid orbits, even "circular" and with a slight inclination, could become "chaotic," due to the "sling" effect of Jupiter's gravity, pick up energy, and turn into cometary orbits. (2) The destruction of small or larger bodies upon atmospheric entry depends in part on the speed of impact in the atmosphere. Such an effect would preferentially select for asteroidal orbits rather than cometary orbits, which correspond to far faster-moving bodies. (3) The survival of these objects when they enter the atmosphere is believed also to depend on their friability. Friable meteorites are destroyed more easily than micrometeorites of the same type, because they slow down at lower altitudes. (4) Certain asteroids are thought to be cometary nuclei that are now "extinct." (5) Bands of interplanetary dust of asteroidal origin have been observed by the infrared satellite IRAS. No doubt they contribute to supplying the micrometeorite flux with particles from asteroids.

One of the most urgent issues is to determine definite criteria for identifying cometary objects in meteorite and micrometeorite collections that are probably rare. It also seems important to understand better the differences between meteorites and micrometeorites and to look for other solar system objects among the micrometeorites. We must also do better in modeling atmospheric transmission of the small and large bodies which bombard the Earth, and calculate the proportion of asteroidal and cometary objects that may intersect the Earth's orbit, utilizing the results of IRAS and the calculations of Jack Wisdom.

THE PASSION FOR PRIMITIVENESS

One of the main concerns of meteoriticists for over twenty years has been to hunt for the most "primitive" material in the solar system. This material would let us go further back, as far back as possible, to the beginnings of the solar system and the formation of our mineral ancestors. Suppose it contained the true original grains swept into swirls like giant nebular cirrus clouds! Then it might give us the first basic materials from which, by modeling, we could "manufacture" the planets, the other less primitive meteorites, and even life on Earth?

However, we have not yet succeeded in reaching complete agreement on this concept of primitiveness. Are the most primitive meteorites necessarily the oldest? We will see that the time interval between the formation of these meteorites and the other ones was very short—less than 20 million years. In this time interval, a meteorite which formed earlier than another could simply have cooled down faster without necessarily being more "primitive" from the standpoint of the characteristics of its initial material. Could it rather have been a meteorite containing the greatest possible concentration of presolar interstellar grains, which would have survived the "remobilization" of the isotopic compositions? Must we sample material from a cometary nucleus by resorting to extremely expensive space missions, assuming that comets are the most primitive objects? Do we already have, without being aware of it, comet samples in our meteorite and micrometeorite collections, as we have lunar and Martian meteorites? These questions drive a highly active and fast-evolving field of research.

The first criterion of primitiveness adopted by meteoriticists requires that the object have a chemical composition similar to that of the atmosphere of the Sun, *considered as a representative sample of the original solar nebula*. Such a conclusion may appear surprising when we remember that the Sun is a sort of nuclear furnace—a huge H bomb, which generates energy equivalent to that released by the explosion of billions of H bombs every single second. How could the composition of the nebula have remained unchanged in this furnace for 4.5 billion years?

Actually, only the interior of the Sun is at sufficiently high temperatures (about 10 million degrees C) to trigger the nuclear reactions that could modify the original chemical composition. In a star with a mass comparable to that of the Sun, this interior does not become mixed with the more superficial and far "cooler" layers (a mere 5,000°C [9,000°F]!) to alter this composition. The light of the Sun comes to us from elements trapped in these layers for the past 4.5 billion years. It is analyzed by techniques of "optical spectroscopy" associated with solar telescopes, and the chemical composition of the Sun's atmosphere, the relic of the nebula, is deduced from them.

Once the composition of this atmosphere had been compared to that of meteorites, it was noticed that the carbonaceous meteorites of the CI type, which are very rare (less than 1% of meteorite falls), of which the one found in the 19th century near the village of Orgueil in France remains the most famous, are the closest to it: the differences do not exceed 10%, for the great majority of the ninety-two natural elements.* In this "classification,"

* Except for the most volatile elements (hydrogen, helium, carbon, and nitrogen) which were lost when the planetesimals were formed, and lithium, beryllium, and boron, which were destroyed by nuclear reactions in the external layers of the Sun.

the other carbonaceous meteorites (CM, CO, CV) are right behind the CIs. All the other ordinary chondritic meteorites are poor in volatile elements by comparison with the CIs, and this depletion increases proportionately to their "metamorphic" index. But it is clear that the parent bodies of these meteorites did not melt completely. Such melting would have produced highly differentiated compositions as they are observed on Earth, Venus, Mars, and the Moon—very different from the "chondritics."

This "chemical" primitiveness of the Orgueil meteorite has been used to determine, with the precision of laboratory analyses, the chemical abundance of the elements—the "cosmic abundances"—which serves as a reference in numerous fields of astrophysics. But we cannot say that this primitiveness is "perfect," as demonstrated by the work of the American mineralogists Harry McSween and Michael Zolensky, because it is limited to chemical composition and does not extend to mineralogical composition. Of the carbonaceous meteorites, the CIs have indices proving that their constituent minerals were completely remobilized by the corrosive action of the water that must initially have existed on their parent bodies, which were still quite "wet" due to the accretion. By now, a meteorite like Orgueil contains practically no more "dry" (anhydrous) minerals, but abundant varieties of extraterrestrial clays which contain a great deal of water in their structure. Moreover, the microfractures that riddle the meteorite have been plugged with veins of water-soluble salts (carbonates and sulfates), suggesting that they had been penetrated by salt-laden water. So the Orgueil meteorite cannot be called truly primitive from a mineralogical standpoint: its minerals were probably not formed in the solar nebula but resulted from a "secondary" aqueous alteration on a parent body. During this alteration, the overall chemical composition of the primitive material did not change. All the elements remained in place, despite the devastating effect of the water on the minerals. And yet the Orgueil meteorite has retained its primitiveness in chemical terms!

This interpretation is not objection-free: the aqueous alteration may have occurred in the nebula before the grains coagulated into planetesimals due to the action of the water vapor present in abundance, which is even more corrosive than liquid water. According to this theory, the alteration was primary and the Orgueil meteorite was even more primitive. Present-day thinking holds that, of the carbonaceous meteorites, those of the CV group—almost as rare as those of the CI group—are the most primitive: they contain neither hydrated minerals nor salt veins. Moreover, their apparent poverty in volatile elements may reflect nothing more than the existence of a mixture of two grain components: one, refractory (chondrules, refractory inclusions), poor in volatile components, which are not found in the Orgueil meteorite; and the other, composed of a fine-grained matrix (less abundant than in the Orgueil meteorite), rich in volatile elements. Primitiveness would result simply from random mixing of these two components in the solar nebula during coagulation of the planetesimals.

The second criterion for primitiveness, still adopted by the majority of meteoriticists, requires that the assembly of minerals be "unequilibrated." On the contrary, in ordinary equilibrated chondrites, which represent about 80% of meteorite falls, the minerals are now equilibrated. Their constituent olivines or pyroxenes have the same composition.

The tiny stratospheric micrometeorites not exceeding one hundredth of a millimeter in diameter are also composed of highly unequilibrated assemblages of minerals. We extended this conclusion to about 80% of the "giant" micrometeorites (about a thousand times the size) collected in Antarctica. This state of disequilibrium indicates the similarity of the small and large meteorites to the primitive meteorites. While there are obvious differences between these three types of extraterrestrial materials within the general framework of this relationship, we have not yet succeeded in understanding them despite our efforts.

The third and more recently defined criterion is that there are small concentrations of presolar interstellar grains in meteorites. According to this criterion, the meteorites and micrometeorites richest in interstellar grains are the most primitive. It is believed that such richness may indicate that their component grains were protected best from the action of as yet poorly explained processes,* which had the effect of profoundly altering the initial isotopic composition of the grains, destroying their primitiveness. At any rate, this is the opinion of Glen Huss, according to whom interstellar grains could easily have been destroyed by the "gentle" metamorphism undergone by ordinary chondrites in the planetesimals, at the same time that it equilibrated their mineralogical composition. Robert Walker's group, however, does not share this point of view. In the primitive meteorites, interstellar grains are different from the other solar components. Hence it is the randomness of turbulence or distance from the Sun, says this group of researchers, that makes a meteorite contain more or fewer interstellar grains. Whatever the case may be, meteorites already classified as primitive according to the other criteria appear to contain more interstellar grains. But within this family, which represents no more than 4% of meteorite falls, no systematic trend can be discerned showing that the CI group, for example, is richer than the other groups.

One point of discussion has recently been raised: that of the primitiveness of cometary grains. Until 1990, it was assumed that comet material was the most primitive, which agreed fairly closely with theories on the formation of the comets, far from the heat of the Sun, in the frost zone—that kingdom of shadow, ice, and snow where primitive material has been preserved by low-

* Probably related to powerful electromagnetic "storms."

temperature freezing. The *in situ* analyses of grains of Halley's Comet in March 1986 by mass spectrometers aboard the Giotto and Vega probes appeared to confirm the greater primitiveness of the comet. The high concentration of organic grains, called "CHON" to denote their richness in carbon (C), hydrogen (H), oxygen (O), and nitrogen (N), was pointed out in particular.

Harry McSween was the first to challenge this persistent dogma by presenting a long list of processes that could have altered the composition of grains trapped in the top few meters of the "dirty" ice constituting the cometary nucleus. Sublimation of this ice near the Sun feeds the jets of gas and dust emitted by Halley's Comet. Thus, one of these processes is the "aqueous alteration" already suggested in the case of the Orgueil meteorite because water vapor, which is corrosive, is the most abundant gas released by this comet and simultaneously entrains the grains.

More recently, geochemists like the Australian Ross S. Taylor* have actually reviewed the analyses of these grains and propose a very different interpretation, which could once more upset our concepts of primitiveness. For example, you choose three abundant elements that can be measured with reasonable accuracy in Halley's Comet: aluminum (Al), silicon (Si), and iron (Fe). You plot a simple geochemical diagram with the ratio between alu-

---

* Already known for his numerous contributions to the problem of the origin of the Earth and the Moon, and with whom I have enjoyed highly stimulating conversations.

minum and silicon concentrations on one axis and the iron-silicon ratio on the other axis, determined not only for the Halley's Comet grains but also for the solar atmosphere and a whole variety of meteorites, including those believed to be the most "primitive" in chemical terms: the CI group. When you do this, you arrive at clusters of points (one point per analysis). Surprisingly, you find that the cluster defined by the points for Halley's Comet is very much to the left of that defined by the CI group, the Sun, and the other meteorites. This seems to shift cometary material from the "chemical" primitiveness that was established relative to the composition of the solar atmosphere and the CI meteorites. If indeed the comet grains were the most primitive, the carbonaceous meteorites and the atmosphere of the Sun should be considered to be differentiated objects. This appears to be unacceptable … just now, at least!

So we have been unable to give a definite answer to the burning question of what is the most primitive object in the solar system. We have to confirm the differences between comets and primitive meteorites, and find criteria to identify comet samples which may be present in the collections of "unequilibrated" micrometeorites and meteorites. By analogy with the primitive meteorites, it could be that the solid grains scattered in the dirty ice of the core of Halley's Comet results from a mixture of components, some of which are truly primitive while others lost their primitiveness in the course of as yet mysterious nebular processes. If these differences are confirmed, it will be even more desirable to send up a space mission that would pick up a comet sample automatically. Are there chondrules and refractory inclusions in cometary ice in addition to a presolar grain component which we assume and hope will be abundant and varied?

HUNTING FOR ISOTOPES

It is time for us to show why *isotopes* are one of the essential tools for our studies. On Earth, as in meteorites, there are ninety-two natural elements characterized by specific chemical properties. You might expect these ninety-two elements to correspond to ninety-two atoms whose mass would increase from hydrogen, the lightest, to uranium, the heaviest. But the situation turns out to be far more complicated. In fact, they correspond to over 400 atoms because, for any given chemical element, there are several atoms with different atomic masses; they have the same chemical properties but different nuclear properties, and constitute the different "isotopes" of the element (the mass of an atom depends on the number of protons and neutrons in its nucleus, with the number of protons being fixed). The unit of atomic mass ($M = 1$) is that of the lightest atom, hydrogen, whose nucleus consists of a single proton. But there are other "varieties" of hydrogen obtained by adding neutrons to the nucleus: these are deuterium (^2H) or "heavy hydrogen" with a mass of 2, and tritium (^3H) with a mass of 3.

When any element is isolated by successive purification steps, what we actually obtain is a mixture of the isotopes of this element. The proportion of each isotope in this mixture defines the "isotopic" composition of the element. There are sometimes "anomalies"—of which meteoriticists are extremely fond—in the isotopic composition of the elements of meteorites, meaning marked deviations in the isotopic compositions relative to those measured on natural reference samples (such as sea water in the case of oxygen), deviations which cannot be accounted for by known processes such as melting or vaporization. These isotopic compositions, very difficult to measure with any accuracy, thus play the same role as fossils in prehistory: they are an essential tool in the attempt to retrace the galactic adventure of the formation of the solar system and the planets, and hence our presolar past.

So that we can understand the magic of isotopes, let us take a look at terrestrial uranium, with its chemical symbol U, which can be extracted from uranium ores. This element, used to fuel numerous nuclear power plants, is composed of two isotopes, one with a mass of 235 (uranium-235) and the other with a mass of 238 (uranium-238). These isotopes differ only by their mass and their very different nuclear properties, since uranium-235 is a far better fuel for nuclear power plants than uranium-238. But their chemical properties are strictly identical, meaning that uranium-235 cannot be separated from uranium-238 in the purely chemical treatment of the ore which yields the uranium "metal" where the concentration of uranium-235 (about 1%) is far less than that of uranium-238. The two isotopes can only be separated by physical treatments repeated over and over again such as gaseous diffusion, liquid ultracentrifuging, or magnetic separation of a jet of ionized atoms using a mass spectrometer (that is, a type of electromagnetic balance), whereby natural uranium is enriched with uranium-235 so that it can be used as a nuclear fuel. The isotopic composition of this enriched uranium will of course appear to be "anomalously" high by comparison with that of natural uranium. But once it has been "burned" in the nuclear furnace of the reactor, this composition becomes "anomalously" low. The isotopic composition of an element will thus depend not only on the way in which it was made but also on the "nuclear" and "physical" history of its own life. Here we find a sort of complex and very powerful fingerprint which meteoriticists have put to a great deal of use.

We will discuss mainly the isotopic composition of the light elements—known as "biogenic" elements because they are the building blocks of the living cell—such as hydrogen, carbon, nitrogen, and oxygen, then the heavier elements such as aluminum, magnesium, rare (or noble) gases,* uranium, and lead. On Earth, the isotopic composition of hydrogen, defined by the ratio of the concentrations of ^{1}H to ^{2}H in sea water, is approximately

* The rare gases, which are chemically inert, include helium, neon, argon, krypton, xenon, and (radioactive) radon.

5,000 to 1. For carbon, the isotopic composition measured, for example, in chalk comprises three atoms with masses of 12 (carbon-12, the most abundant isotope on Earth), 13, and 14 (carbon-13 and carbon-14); the ratio of carbon-12 to carbon-13 is approximately 90 to 1 on Earth while it may exceed 5,000 to 1 in the envelopes of certain stars. The oxygen that we breathe has three isotopes: oxygen-16, the most abundant, oxygen-17, the rarest, and oxygen-18. Of the rare gases, we will use neon, with three isotopes (neon-20, neon-21, and neon-22), and xenon, which beats all records with nine stable isotopes (with masses between 124 and 136) and which for this very reason has been put to all kinds of uses. Lead (Pb) has four isotopes (lead-208, lead-207, lead-206, and lead-204), the most abundant of which is lead-204.

Elements are formed by various nucleosynthesis processes in the stars, which can be thought of as gigantic nuclear caldrons. At certain times in their lives, stars went through various cataclysms (stellar "winds" and eruptions, novas, and supernovas) during which huge gusts of gas were hurled out from their atmospheres into interstellar space. As these gusts cooled, grains condensed, trapping various isotopes synthesized by the parent star. Thus they have very characteristic isotopic "fingerprints." In interstellar space they became mixed with other grains, synthesized in other stars, at other epochs, with other isotopic composition. Then they became incorporated into a huge interstellar cloud at least 100,000 times the mass of the Sun, of which a tiny fragment produced the solar nebula. Contraction of this nebula by gravity led to the formation of meteorites and planets. Before the discovery of isotopic "anomalies," meteoriticists considered that the presolar interstellar grains had lost their isotopic imprint in the chaotic tumult and intense heat that accompanied the birth of the Sun and the planets, which caused them to heat up enormously or evaporate, so that their isotopic composition became equilibrated. But while a few grains may have escaped the nebular "fire," their incorporation into planetesimals, compacted, heated, and fractured by collisions before the meteorites escaped and fell to Earth, would have been sufficient to wipe out any presolar isotopic fingerprint.

At the time they were produced in the stars, many isotopes were radioactive (unstable), and decayed into stable daughter elements, often the isotopes of other elements, with different chemical properties than the parent. Radioactive decay occurs at a constant rate, thus defining a sort of life expectancy for the isotope, called "half-life τ," so that after an interval of time equal to τ, half the atoms have disappeared.* If τ is long, comparable to the age of the solar system (approximately 4.56 billion years ago), they are still observed today in meteorites and terrestrial samples. If not, when τ is too short (less than a

* For uranium-235, the half-life is about 700 million years. If at time $t = 0$ we have 1,000 atoms of uranium-235, we will have 500 after 700 million years, 250 after 1,400 million years, etc. Today there is one-eightieth the amount of uranium-235 there was at the beginning of the solar system (believed to be 4.56 billion years ago).

hundred million years), their concentrations become too small to be measurable. The radioactive ash of the stars is then said to be "extinct."

MODEL AGES AND FORMATION INTERVALS

A pair of isotopes composed of a "parent" (radioactive) and a "daughter" can be used like a chronometer to measure the ages of different types of substances.

When the half-lives are long (τ equal to approximately one billion years), we still detect the presence of parents in the minerals of the meteorites. When we measure the concentration of both parents and daughters, we can determine the *ages of formation* of these minerals and date the time at which the parent-daughter pair was hermetically trapped because cooling of the meteorite reached a sufficiently low temperature for the isotopes to be unable to separate from each other, migrate, and "equilibrate" themselves with the isotopes around them.

The rubidium-strontium method is based on decay of the radioactive isotope of rubidium which has a mass of 87 (τ approximately 49 billion years) to give the isotope of strontium, with a mass of 87. By this method, which is the most widely used, the team of Jerry Wasserburg (with which that of Claude Allègre has been successfully rivaling since 1975) was able to determine the first precise ages of meteorites in 1967. The method indicates that the oldest terrestrial rocks (–3.8 billion years) and lunar rocks (–4.2 billion years) are younger than most meteorites.

When in 1960 the American Clair Patterson evaluated the age of the first meteorites at 4.55 billion years ago, he used uranium-235 (τ approximately 700 million years) and uranium-238 (τ approximately 4 billion years) which decay into lead isotopes with masses of 207 and 206. Like any age, however, this is a "model" age: its determination depends on the assumed value of the initial isotopic compositions of lead and uranium just before they were trapped in the minerals of the meteorites. In this way, Patterson determined that the (model) age of the Earth was similar to that of the meteorites because he assumed that the initial isotopic composition of lead on Earth was identical to that measured for the sulfides in the Canyon Diablo iron meteorite. Today, some research teams are able to measure the age of meteorites with a precision of within 1 million years.

But when we consider these isotopes with long half-lives, still present in meteorites, it is still impossible to know whether this age of the Earth and the meteorites is actually that of the solar nebula or the Sun. After all, meteorites and the Earth may have formed a billion years after the Sun. To answer this crucial question, meteoriticists figured out how to find and use other parent radioactive isotopes with far shorter half-lives, approximately one to a hundred million years, which are now "extinct." The difference between the age of the (extinct) parent and the daughter is interpreted as a *formation interval*. If the daughter of one parent is found in meteorites, this implies that the time intervals between nucleosynthesis of the parent in a star, its incorporation into the solar nebula, then its trapping in the minerals of meteorites has not greatly exceeded its half-life; otherwise it would have disappeared before being able to scatter about its daughters in any appreciable quantities.

Several chronometers are available for evaluating these intervals. The first, proposed by John Reynold's team in 1960, is the "xenon-iodine" method based on the observation of an "excess" of xenon-129 (which had originally been considered an isotopic "anomaly"), produced by the decay of a radioactive isotope of iodine, iodine-129 (τ approximately 16 million years). This yielded an initial valuable piece of information: most meteorites contain roughly (within a factor of 2) the same "excess" of xenon-129. We may conclude that no more than 20 million years could have elapsed between the first and last meteorites formed. What is more, using a model of chemical evolution of the elements, yielding the initial concentration of iodine-129, we have been able to show that the formation interval between the end of iodine-129 nucleosynthesis and its incorporation into the "latest" meteorite is approximately a hundred million years.

This interval is longer than the interval (a few million years) deduced by Jerry Wasserburg's team in 1976 for the radioactive isotope of aluminum (aluminum-26), which decays into a magnesium (magnesium-26) with a half-life of 700,000 years. Although we still do not quite understand the reasons for this wide gap between formation intervals when deduced from extinct aluminum or iodine, it can be concluded that the formation of the solar nebula, and that of the Sun, the meteorites, and the planets, occurred in very rapid succession over less than 100 million years, after the last stellar nucleosynthesis responsible for this injection of "last-minute" isotopes. Hence we can say that all the objects in the solar system were formed "simultaneously."

ISOTOPIC ANOMALIES BEFORE THE ION MICROPROBE ERA

In 1973, the American Robert Clayton and his associates in their turn revolutionized our knowledge without resorting to radioactive isotopes. When they analyzed refractory inclusions in the Allende (Mexico) meteorite, which weighed a few milligrams, they discovered anomalies in the composition of the three stable isotopes of oxygen. These inclusions contained too much oxygen-16 by comparison with the measurements made on innumerable terrestrial samples. These excesses cannot be interpreted through processes known on Earth, such as melting or vaporization, as these would very slightly decrease the amount of oxygen-16 contained in the residual solid. When this analysis was extended to many meteorites, it revealed wide variations in this excess of oxygen-16, not only between one group of meteorites and another, *but also between the components of one and the same primitive meteorite*, such as "dry," anhydrous minerals, the chondrules, and the fine-grained matrix.

No longer was it possible for the families of meteorites to be derived from each other by known processes. For better (or worse), Clayton's discovery destroyed overnight the dogma of isotopic homogeneity of the solar nebula which implied, in particular, that all the grains of interstellar dust initially present in this nebula had been destroyed, yielding new grains that were "homogeneous" from the standpoint of their isotopic composition.

In the asteroid belt, the excess oxygen-16 found in the various classes of meteorites (thought to be derived from the asteroids) appears to have varied over distances on the order of a few tens of astronomical units. Such distances correspond roughly to those separating the asteroids that could have produced the various groups of meteorites classified according to their excess oxygen-16. Today, Clayton believes that these variations in the isotopic composition of oxygen result from exchange between the oxygen in the gas of the solar nebula, which has a thoroughly homogenized isotopic concentration, and that of the highly heterogeneous and highly unequilibrated presolar grains formed at different epochs and in different places. This exchange, activated by as yet poorly understood intense local heating ("hot" spots), would have equilibrated the isotopic composition of the grains more or less completely.

This discovery was followed up in 1987 by the observation of isotopic anomalies affecting other elements; these anomalies, still more spectacular, allow "presolar" interstellar grains to be identified in the primitive meteorites. Once more, these discoveries had their roots in studies of rare gases. In 1969, Dave Black and Robert Pepin discovered an excess of neon-22 in primitive meteorites. This phenomenon could only be explained by resorting to theories of nucleosynthesis, which implied that this isotope must have derived from the decay of a sodium isotope (Na) with a mass of 22, which could only form when a star exploded to turn into a nova. It had to be assumed that this sodium and this neon had been incorporated into interstellar grains formed by some condensation in the expanding envelope of the same star. These grains had miraculously survived the heating of the solar nebula.

In 1976, we attempted (unsuccessfully, with the limited techniques of the time) to identify the presolar grains containing these anomalies in order to characterize them. We used a "concentrate"* of minerals, obtained by Peter Eberhardt at the University of Bern from about one gram of the Orgueil meteorite in which he measured a very large excess of neon-22.

In 1985, the American Edward Anders and his associates looked at the decay products of "superheavy" elements (heavier than uranium) whose existence had been postulated. When they analyzed the isotopic composition of xenon, they found a slight anomaly which had never previously been described. To identify the host phase of this anomaly, they developed a very sophisti-

* This concentrate had been prepared by "physical" separation methods such as sedimentation in liquids of different densities.

cated technique, more efficient than Eberhardt's: they took about 30 grams of meteorites which they subjected for *two months* to attack in very strong chemical reagents, and collected a few tiny "etching residues" weighing a few milligrams. They discovered that these residues were very rich in tiny diamonds, with diameters of a few hundredths of a thousandth of a millimeter, and in "larger" grains (about a thousandth of a millimeter) made of silicon carbide and a sort of graphite. They quickly deduced that these were interstellar presolar grains, because these residues contained all the abnormal excess xenon.

These "anomalous" grains represent a small fraction—less than one thousandth—of the normal constituent grains of meteorites, whose isotopic compositions are equilibrated. Just like the grains rich in neon-22 or oxygen-16, they managed to escape very intense local heating which wiped out any presolar signature from the majority of the grains which constitute a meteorite today, particularly by equilibrating the isotopic ratios. As in the case of Clayton's oxygen, the word "anomalies" means that they cannot be obtained by modifying terrestrial compositions by any known processes.

Shortly thereafter, the grains ushered in the era of multidisciplinary analytical microscopy, which is based in large part on the ion microprobe of the French physicist Georges Slodzian.

THE MICROSCOPIC HUNT FOR PRESOLAR GRAINS

The team of Robert M. Walker, who had been polishing up his microanalysis tools for nearly five years, then took over the new challenge of studying in far greater detail, *grain by grain*, the famous etching residues. One spectacular discovery would follow another, as we shall see when we talk about one of the most revolutionary aspects of modern meteorite studies.

In 1978, Walker had predicted the important role that Georges Slodzian's ion microprobe would play in this type of research. He tried to put together the funding needed to purchase the first ion probe marketed by a French company. At the same time he hired an Austrian physicist, Ernst Zinner, to work on the instrument itself —in fact, he made a number of decisive improvements. In this instrument, a very narrow beam of so-called "primary" ions bombards the surface to be analyzed, pulverizing the atoms constituting its surface. These atoms are then separated magnetically through a mass spectrometer. From the mass spectrum of the atoms one can deduce both the chemical composition* of the tiny grains and their isotopic composition of hydrogen, carbon, nitrogen, oxygen, and other elements except for the rare gases which still require highly specific mass spectrometers.

* Including not only the major elements but also the "minor" and "trace" elements present in very low concentrations, less than 1% and 0.01% respectively.

The extreme sensitivity of this technique represents a major step forward, as one can work on tiny individual grains measuring about a thousandth of a millimeter, not on samples a billion times heavier as in the technique used by Anders and Clayton which can determine only the average chemical or isotopic compositions of millions of grains. The group simultaneously developed other "analytical microscopy" techniques, preparing for multidisciplinary microanalysis of any tiny grain with "exotic" characteristics.

In 1988, after a great deal of laborious work to improve the techniques, discoveries began to flood in. I cannot list every one of them: we will confine ourselves to grains of silicon carbide, with the chemical formula SiC. It was shown, for example, that in the individual grains of silicon carbide isolated by the Anders team, the scatter in the values of carbon-12 to carbon-13 ratios, which range from about 3 to over 3,000 (terrestrial value about 90) corresponds to that observed in the envelopes of a particular type of star, the red giants belonging to the carbon-rich *asymptotic giant branch* (AGB).

In the last few years, Robert Nichols, on the same team, has focused a very narrow laser beam on *individual* grains of silicon carbide, thus releasing the helium and neon trapped in the grains, and analyzed them with an ultrasensitive rare-gas spectrometer, probably the world champion in its class. In this way he showed that about 5% of these grains had anomalies in neon composition; synthetic anomalies could be found only in AGB stars.

The ion microprobe also confirms that they embody other isotopic anomalies, observed not only for the major elements (carbon and silicon) but also for minor and trace elements (nitrogen, magnesium, titanium, barium, and neodymium). A subfamily of silicon carbide grains, called X grains, is thought even to have isotopic anomalies which would suggest that they originated in another stellar cataclysm, a supernova. These isotopic compositions are completely different from the values found for any other object in the solar system. The presolar origin of the grains is no longer in doubt: they were formed when gases were ejected from the atmospheres of the stars responsible for synthesizing the isotopes, cooled down, and condensed. But despite uncertainties as to the number and type of stars participating in synthesis of these isotopes, the interstellar presolar grains have shown up at the right time and in the right place.

The possibilities of the ion probes appear to be limitless. Scientists have even found, inside silicon carbide grains, high concentrations of magnesium-26, the daughter of aluminum-26. Once again this implies a very brief formation interval for the solar system, but this time it is determined for *individual* grains measuring a few thousandths of a millimeter. When we combine this probe with other microanalytical instruments such as the electron microscope, we can peer into the innards of just one of these grains in still further detail. Thomas Bernatowitcz, in the same group, succeeded in finding, inside an interstellar graphite grain measuring six thousandths of a millimeter, a minuscule interstellar grain of titanium carbide, about one-hundredth the size, which itself will probably present anomalies in the isotopic composition of the carbon and titanium when analyzed with the ion probe!

Analysis of these anomalies grain by grain should enable us to identify our parent stars among the red giants, novas, supernovas, and other exotic objects (Wolf-Rayet stars, Bok globules, etc.) and specify some characteristics of the interstellar cloud from which the solar nebula originated. A new and exciting phase of modern astronomy has just opened, that of the presolar interstellar dust (PID) where the ion microprobe plays the role of a telescope and where meteoriticists must henceforth walk hand in hand with astronomers who use infrared astronomy and millimeter astronomy to penetrate the cocoons where the young stars form. They will also have to cooperate with specialists in nucleosynthesis who establish theories of isotopic anomalies. So this new "PID" astronomy is indissociable from other forms of astronomy.

This reign of microanalysis is only just beginning. For the time being it is the prerogative of the Americans, particularly Robert Walker's laboratory at Washington University in St. Louis, Missouri. France could have been in the forefront of this modern research, since the French invented most of the microanalytical techniques used in this "PID" astronomy. With a little luck in 1976, and a few more resources, we probably would have been the discoverers of the interstellar grains. But it was not to be: we were up against powerful lobbies of astronomers, planetologists, and geochemists who had other priorities and were more skilled in wangling funds for sometimes less interesting research. Moreover, the CNES (French National Space Research Center) adopted a hands-off policy by refusing thus far to support directly the preparation of the return to Earth of cometary and Martian samples, in which very fine dust will probably be *the most interesting* main component. NASA was farsighted enough to support such preparation starting in 1980 (even though the mission scheduled for this purpose, "CRAFT," has just been scrubbed!) because the cost of this preparation is very small when compared to the cost of the mission. Without good, motivated teams and without constantly improved microanalytical techniques, it seems pointless to engage in a very expensive space mission. The Americans will be far better than the French in exploiting the results.

IV

FROM THE PRESOLAR TIME SCALE TO THE ORIGIN OF LIFE

THE PRESOLAR TIME SCALE OF CATACLYSMS

Recent studies by astronomers, astrophysicists, and meteoriticists suggest that there are two cosmic time scales. The first covers our presolar history, those several hundreds of millions of years before the formation of the Sun which rolled by in a gigantic cloud of interstellar gas and dust. The second, a "nebular" time scale, covers a period of about a hundred million years during which the Sun, the planets, their satellites, and the smaller bodies of the solar system were formed by the action of a giant gas bubble given off by our last parent star as it "died."

A decade ago, little attention was given to our presolar history because there were insufficient data to decipher it. The situation changed radically, as we have seen, with the discovery of isotopic anomalies and presolar grains in meteorites, then with the great strides made by astronomers and astrophysicists over the last decade.

Those who specialize in nucleosynthesis improved on certain calculations used to "theoretically" fabricate the chemical elements in the gigantic cataclysms brought about by explosions and "eruptions" in stars of different types. This synthesis produced the heavy elements to make the grains of dust essential for the building of the planets, as well as the radioactive isotopes, extinct today, which spiced the meteorites with isotopic anomalies. Using both these anomalies and explosive nucleosynthesis calculations, we obtain various "formation intervals" for the elements that let us establish the dates on the presolar time scale. These dates are expressed in "negative" time intervals, $-\Delta T$, measured back from the age of the formation of the solar system, $T_0 = 0$. The sign – simply indicates that we are going back in time, starting from T_0.

It was the astrophysicist Alister G. W. Cameron, himself a specialist in nucleosynthesis, who did the best job of using these data in the presolar history model he proposed in 1991. This model involved a complex succession of "epidemics" of star formation, and we will use it to deduce our "oversimplified" time scale.* Some of the events and dates in it will certainly be questioned in the future; hence, it may be considered plausible, which does not mean it is the most probable. However, it is no science fiction.

* Cameron's model is described in a long and highly technical article, which is impossible to summarize in all its complexity in two pages.

$-\Delta T \approx 130$ MILLION YEARS: BIRTH OF OUR GREAT-GRANDPARENTS

Everything begins in a giant cloud of interstellar dust and gas containing a mass 100,000 times the mass of the Sun (M_S), which measures some 100 million astronomical units. This mass is composed essentially of a gaseous mixture of hydrogen molecules (78% by mass), helium (20%), and heavier atoms (2%) in which 1% of very fine dust grains is scattered. The concentration of hydrogen is approximately 1,000 molecules per cubic centimeter, corresponding to a gas pressure about a hundred quadrillionth that on Earth at sea level.

This cloud, which cannot survive more than 200 million years before it disaggregates, is inhomogeneous. It has large gaseous "cores" whose masses and densities can reach approximately 10,000 M_S and 100,000 molecules per cubic centimeter, respectively. These are propitious conditions for triggering their gravitational collapse, leading to "condensations" of stars. In each core, stars are thus born "simultaneously," in clusters. The most visible and turbulent ones, those that will be the most efficient nucleosynthesizers of heavy elements, are the hottest O- and B-type stars.

Among the OB stars, those with a mass approximately ten times that of the Sun have a very short lifetime because of their high mass. They burn up all their nuclear fuel in less than 10 million years. Then they explode, resulting in the supernova phenomenon in which a single star suddenly shines brighter than billions of stars in our galaxy. A cataclysm of unimaginable power, on the order of a nonillion H bombs, during which a first "furnace batch" of very heavy elements is synthesized.

This explosion blows out the layers of the star, which then propagate initially at very high velocity (approximately 10,000 km/sec or 6,200 mi/sec), causing formation of an interstellar "superbubble" which will locally sweep through the core of the interstellar cloud, "cleaning out" its gas and dust. We can see that the formation of another few supernovas will limit the lifetime of these interstellar cloud cores to about a hundred million years.

But during this interval of about a hundred million years, all the stars in a given core (which are not only O and B types) contribute in their own way to synthesizing heavy stable elements and radioactive elements in the course of different stellar cataclysms, less violent but far more frequent than a supernova. One of the most effective, which we shall be looking at later, involves gigantic and frequent flashes of helium, the nuclear fuel in the AGB stars. Moreover, this type of star ejects a fraction of its surface layer (contaminated by the new synthesized elements) into the surrounding medium every time there is an eruption or emits puffs of its atmosphere in the form of a stellar "wind," racing at speeds between a few tens and a few hundreds of kilometers per second. In young stars, such winds may carry away up to one hundred thousandth of the mass of the star per year. We will also see that certain stars, because of various emissions of matter, themselves start up small, local "epidemics" of star formations.

–ΔT ≈ 25 MILLION YEARS:
BIRTH OF OUR GRANDPARENTS

At –ΔT approximately equal to 25 million years, the first "local"* core of the interstellar cloud has broken up. But a massive new core is reforming where the entire preceding evolution (dominated by associations of O and B stars) starts up again from a material far richer in heavy elements. However, in this new cycle of star formation, which should last another several hundred million years, an event of paramount importance for humankind will occur at –ΔT equal to approximately 2 million years.

–ΔT ≈ 2 MILLION YEARS:
OUR AGB PARENT DIES,
BLOWING OUT A GIGANTIC BUBBLE

Toward the end of their lives, stars swell up hugely into "red giants." Then, depending on their mass, they "die" in one of two main ways. As we have seen, the most massive explode (supernovas), but those whose masses are equal to one or two solar masses turn into AGB stars. After they have burned their way normally through all their hydrogen and helium, they are left with a core of oxygen and carbon, surrounded by a layer of helium, itself covered by a layer of residual hydrogen. This "nuclear sandwich" of hydrogen and helium gives rise to complex phenomena. In brief, the helium and

* "Local" means that the position of the future solar system is in this core.

hydrogen will flare up again, but alternately—never together. So there are sporadic helium flashes (they cannot be seen at the surface of the star because the helium layers are far too deep down). Cameron and others proposed that these flashes are very efficient "furnaces" for nucleosynthesis of heavy elements. Moreover, at the end of their lives, instead of "going supernova" and dying, the AGB stars explode far less cataclysmically, blowing out their hydrogen envelope in the form of a "planetary nebula" with a mass about one-fifth that of the Sun. They then become white dwarfs.

According to Cameron, the gas of this giant bubble divides into two "nebulosities," which move in opposite directions under the influence of the galactic magnetic field. They compress the interstellar medium and, by a "piston" effect, accumulate along their two wakes two small dense cores of interstellar cloud (a few times the mass of the Sun, at most). Each of these cores metamorphoses rapidly into two young stars surrounded by a cocoon of dust. The local epidemic continues to spread: each of these two stars, by means of its strong stellar wind, produces a new bubble which behaves like the planetary nebula of the original AGB star, each giving birth to two new young stars, etc.

Thus at $-\Delta T \approx 2$ million years, a very special AGB star, our parent, has blown off its last bubble, triggering a very localized epidemic of stars. In this epidemic will appear, two million years later, a vague glow, our Sun, spiced with "fresh" elements synthesized by our parent.

EVIDENCE FOR DIAGNOSING A STELLAR EPIDEMIC

It is only in the last decade that we have been talking about this idea of a complicated stellar epidemic producing a cascade of generations of successive stars in less than two hundred million years. Until then, we were content to put forward a single supernova to account for the formation of the solar nebula. Since then, however, ten or so "extinguished" radioactivities have been found in meteorites, as well as presolar grains showing enormous isotopic anomalies. It is not possible to synthesize such a variety of products with a single star.

At the same time, improvements have been made in calculations of explosive nucleosynthesis and in the models describing the progressive enrichment of the interstellar medium with heavy elements thus synthesized. It was clear from these calculations that the original supernovas did not enrich the local interstellar medium sufficiently with the heavy elements (from oxygen to iron) necessary to bring about a sufficient concentration of dust grains in the solar nebula: the processes of grain "coagulation" would have been too slow and too inefficient. On the other hand, in the Cameron model, starting from the second generation of stars, the local interstellar medium has become sufficiently enriched with these elements, synthesized in the course of events far more frequent and far less cataclysmic than a supernova.

Nucleosynthesis calculations allow the type of star explosions involved in these stellar epidemics to be identified once the chemical and isotopic composition of the individual presolar interstellar dust grains trapped in meteorites has been characterized. Edward Anders, who initiated the discovery of interstellar grains in meteorite etching residues, estimates that it would take probably a thousand stars to account for the diversity of the isotopic anomalies observed with the ion microprobe in the presolar grains of meteorites. This number seems too high to Ernst Zinner and Robert Walker, who collaborated with Anders, and to nucleosynthesis specialists like Donald D. Clayton. It will take another few years before we become more familiar with the number and type of stars in the complex stellar epidemic which gave rise to our most primitive mineral ancestors.

As I write, I have just received a news flash. Until September 1992, we did not know the type of star that synthesized fluorine (used, for example, in toothpaste). Astronomers have just discovered it in AGB stars, which are similar to our putative parent! The existence of these "toothpastes" reinforces my confidence in Cameron.

THE NEBULAR TIME SCALE

The gravitational collapse of the nebula ejected by the AGB star commences. This nebula will become isolated, revolve about itself, and no longer mix its material with that of the far more dilute interstellar medium. The accretion of new interstellar matter (radioactive atoms, interstellar grains, etc.) will become negligible. This is the *instant T = 0 of formation of the solar system.* A new time scale is starting, where we will note the principal events that led to formation of the Sun, the planets, and the parent bodies of the meteorites and micrometeorites. The dates on the time scale are time intervals, ΔT, this time positive, which have elapsed since T = 0. When we accept this convention today, ΔT is more or less equal to 4.56 billion years. This time scale is based on the recent work of several astrophysicists (Alister W. G. Cameron, Gregor E. Morfill, and Stuart J. Weidenshilling) and meteoriticists (Ross S. Taylor, George W. Wetherill, and John A. Wood), who have endeavored to integrate very different kinds of studies in their models.

Paradoxically, we understand this nebula past less well than our presolar history, the scene of gigantic, mind-boggling cataclysms. These cataclysms, albeit rare, can actually be observed by astronomers because of the large number of stars in our galaxy (about 200 billion) and other galaxies. By observing them, we can check the validity of the astrophysicists' models. On the other hand, to establish the nebular time scale, we have no observations on the sequence of complex processes that allowed us to pass from the homogeneous "solar" nebula, several millions of times heavier than Earth, "clogged up" with very fine dust, to the highly heterogeneous distribution of masses currently concentrated in the nine planets, their sixty or so satellites, the comets, and the asteroids.

BETWEEN 10,000 AND 100,000 YEARS: FORMATION OF THE PROTOSUN

The interstellar matter of the nebula will progressively gather at its center, in "free fall," until it initially forms a protostar twenty-five times larger than the present Sun; when the hydrogen begins to burn by nuclear fusion, its brightness reaches a level three times that of our Sun. The nebular disk then settles into the shape proposed by Kant and Laplace, extending up to 100 astronomical units. The ultraviolet radiation of the young Sun produces ions in the gas of the nebula; the interaction between these ions and the strong magnetic field of the Sun slows it down, producing a sort of viscous braking effect. This braking prevents it from revolving 400 times faster than today and from reaching an equatorial velocity greater than the speed of light ... which, as we know, is impossible. This overcomes the main objection which led to the Kant and Laplace model being rejected for nearly 300 years.

Shortly after its formation, the young Sun passes through a phase of evolution similar to that observed in young stars of the T-Tauri and FU-Orionis type, lasting about one to ten million years. These stars eject bursts of their atmosphere in the form of stellar winds, moving at velocities of up to several hundreds of kilometers per second, the intensity of which may be one million times that of the solar wind of today; they give off an enormous amount of ultraviolet radiation. If the Sun went through this stage of evolution, its wind was sufficiently intense to clean out all the gas and dust from the solar nebula. This implies that the settling of dust into sufficiently massive bodies (about 10 m [33 ft] in diameter) to resist this wind occurred before the wind started. If it had not, we would not exist. There is every reason to believe that this intense ultraviolet flux contributed to modifying the primitive atmospheres of the terrestrial (inner) planets. So our earliest nebular history appears to have been strongly dependent on the sudden changes of mood in the young Sun.

IN LESS THAN 100,000 YEARS: METAMORPHOSIS OF INTERSTELLAR GRAINS INTO SOLAR GRAINS. THE CHONDRULE "FACTORY"

As it contracts, the nebula heats up because it is giving off gravitational energy and the young Sun appears; the temperature drops as the distance from the Sun increases, producing a thermal "gradient." Models popular before the discovery of interstellar grains in meteorites acknowledged that the solar nebula had a small mass, about one-hundredth that of the Sun, which extended over thirty astronomical units. It was believed that the temperature there was sufficiently high to volatilize all the interstellar grains up to the asteroid belt, at least. Recondensation of the vapor thus produced, when the nebula cooled down, would then have given the grains a "normal," homogeneous, isotopic composition.

Now we have to explain the isotopic anomalies and survival of a minor fraction of interstellar grains, observed in meteorites. One of Cameron's nebula models, which is more massive and cooler, would better match infrared astronomy observations of young stars. But the temperature would be too low in the asteroid belt to volatilize the presolar grains or melt them locally. So we search for other sources of energy, with limited dimensions, superimposed "locally" on the general thermal gradient, to produce, for example, the heating necessary for the formation of chondrules or the volatilization of grains.

Meteoriticists have shown that chondrules were formed by melting followed by sudden cooling, a sort of "quenching." They constitute one of the major components of nearly 85% of meteorites, and may represent 75% of the total mass of just one of them. A veritable chondrule "factory" existed within the solar nebula.

Various mechanisms have been proposed to account for local hot spots capable of forming chondrules, such as different types of "electromagnetic storms." The astrophysicist Gregor Morfill postulates the existence of giant flashes of lightning in the nebula, caused by electrostatic charging of grains of dust or ice. On Earth, the core of a lightning flash is about one millimeter in diameter, and it lasts one-thousandth of a second. In the nebula, these diameters and times would have been considerably greater: as much as one kilometer and one second, respectively.

The more locally dust-laden the nebula, the greater would be the frequency and intensity of these nebular storms. Chondrules would have been formed during these electromagnetic storms from very fine presolar grains "burned up" by this lightning and turned into stony nebular "hailstones." Areas with a low concentration of presolar dust would escape these storms, allowing interstellar grains to survive and then become mixed with the other "metamorphosed" grains.

One can imagine the nebula at this point as being dotted with huge, highly dilute clouds carrying different types of "nebular hailstones." In these turbulent, lightning-strewn cloud formations, particles winked out and re-formed incessantly. The clouds may have looked like cirrus clouds in the Earth's sky, filmy trails that formed at high altitudes in a very rarefied atmosphere, composed of particles of ice.

IN JUST OVER 100,000 YEARS: "NONGRAVITATIONAL" COAGULATION OF NEBULAR HAILSTONES INTO PLANETESIMALS

Now that we have hailstones and dust dispersed in a gas, they have to come together to form as large a body as the initial rocky core of Jupiter, which has about 15 times the mass of the Earth. The process consists of several stages.

If the Sun went through a T-Tauri phase, formation and accretion of these various components into far more massive bodies must have been rapid; otherwise they would have been rapidly dispersed by the solar wind.

Another reason in favor of this hypothesis is that the various classes of stony meteorites have marked differences, particularly concerning the features of the chondrule families. For these differences to have subsisted, they must not have had the time to be dispersed by the winds and be mixed. It is currently estimated that the width of the "allowed" mixing zones could not have been greater than a few tenths of an astronomical unit.*

Ten or so years ago, it was believed that the problem of accretion had been solved by the "gravitational avalanche" model of Dave Ward and Peter Goldreich. This model described something like giant battles waged by hailstone balls which led to the formation of larger and larger bodies in the nebula. According to this model, the original grains settled toward the middle plane of the nebula. When their number per cubic kilometer exceeded a critical threshold, a sort of gigantic avalanche was spontaneously triggered, during which the grains coagulated into balls as large as a fist. Then a calm set in. The balls then resumed settling until their concentration was once more sufficient to set off another avalanche, producing larger bodies, and so forth.

---

* This width must correspond to that of a zone where a particular family of asteroids clusters, which could be the parent bodies of a given type of meteorite. When you take a "step" of 0.1 astronomical unit in the asteroid belt, the composition of the asteroids varies greatly. The mixing of the various mineral components of a meteorite or an asteroid thus occurred only within this "small" distance of 0.1 astronomical unit (15 million km [9 million mi]!), otherwise we would not have different families of asteroids and meteorites.

But Weidenshilling showed that this mechanism could become effective only if the turbulence (reflecting the velocity of the gas relative to the solid matter in suspension) did not exceed a few meters per year. (Higher velocities would have prevented this progressive and "orderly" settling.) This value, which is far too small, appears to be totally incompatible with observations made on turbulence velocities in cavities of young stars.

So this author called for a turbulence that would be "just right" (not too low and not too high) of 10 to 100 km [6-60 mi] per hour. This is manifested by a sort of nebular wind producing low-speed collisions between the grains, which are actually moving in kinds of Keplerian rings centered on the Sun. If we assume that there is some type of glue on the surface of the grains (for example, a thin film of carbonaceous matter or ice), these collisions may produce coagulation; the grains will gradually grow to a size of 1 to 10 km [0.6-6 mi] in diameter (defining planetesimals), without gravitational attraction having to be involved.

IN LESS THAN 10 MILLION YEARS: GRAVITATIONAL COAGULATION OF PLANETESIMALS INTO PLANETARY EMBRYOS

Once they have reached this size, the planetesimals are no longer swept along by the nebular winds. But the largest of them will grow by accretion as the smallest are attracted to them by gravity and crash into them. According to the more recent models, this process will pick up speed, something like a chain reaction, to the benefit of the most massive bodies. The accretion "reaction" goes faster and faster, producing an avalanche of impacts.

George Wetherill proposed numerical simulations, which are currently the most elaborate for describing this avalanche, confining himself to the formation of the terrestrial planets.* Although the conclusions from these simulations remain uncertain, they are extremely valuable in "visualizing" a highly complex process that takes place in two principal stages lasting very different lengths of time.

To describe the formation of the terrestrial planets, you start from about ten billion planetesimals confined in a sort of gigantic "theoretical" swarm which extends roughly from the position of Mercury (0.4 astronomical unit) to that of Mars (1.5 astronomical unit). Then, in the computer, you "turn on" well-known physical interactions such as gravitational attraction between planetesimals, their braking in the nebular gas, and so on. As time goes by—1,000 years, 10,000 years, 100,000 years, a million years, 10 million years, etc.—the swarm of tiny points evolves continuously on the screen: actually a speeded-up film of this part of our history.**

---

* Even the most powerful computers cannot at the present time describe this process for a planet 300 times the mass of the Earth, involving coagulation of a far larger number of planetesimals dispersed in a huge cavity.

** In fact, the computer program developed by G. Wetherill generates only tediously long columns of numbers on the computer screen. An animated graphics program would have to be written to show the swarm of points coagulating and exploding on the screen.

Now comes the surprise. After only 100,000 years, the swarm has thinned out. You can clearly see half a dozen far larger points resulting from coagulation of small points, and those will give rise to the planets. In fact, in this first and short stage, the swarm of planetesimals coagulates into a few thousand bodies called "planetary embryos," with sizes greater than 1,000 km [600 mi], comparable to or greater than that of Ceres, the largest asteroid. This fast coagulation is accounted for because at the outset the planetesimals were close to each other (about 500 km [300 mi] apart) so that the collisions causing coagulation were frequent. During the second stage, a much slower one, the hundred or so embryos larger than 4,000 km [2,500 mi] will, by giant impacts with the largest ones, give rise to the terrestrial planets with their present characteristics (mass, orbit, speed of rotation).

IN LESS THAN 100 MILLION YEARS: THE PLANETS REACH THEIR FINAL DIMENSIONS

During this second stage, the distances between the embryos are far greater (several million miles), and space continues to empty out, so collisions are far less frequent. Digital simulations indicate that the coagulation time increases, to a range of ten to a hundred million years.

These calculations disclose spectacular scenarios. For example, in *one* of the two hundred or so numerical simulations that have been run so far, Venus and Earth appear simultaneously, by direct coagulation of embryos, with no intermediate stage. Mercury, on the other hand, is formed well before the Earth. But upon a giant impact the planet would be shattered into multiple fragments which would re-accrete to produce eventually "our" planet Mercury, as we know it. In these simulations, Mars also exists before the Earth. But it will eventually be ejected from the solar system. However, such simulation manages to restore only three of the terrestrial planets.

The simulations offering a family of "good" scenarios that would place at least two of the four terrestrial planets in the right spot, with the right mass and the right rotational speed, can be estimated at 10%. But all the simulations agree that the end of the growth of the planets was cataclysmic, because it involved giant impacts with bodies more than 3,500 km [2,200 mi] in size (corresponding to the diameter of the Moon). In particular, the energy released by these collisions, which took place at about 10 km/sec [6 mi/sec], was quite enough to melt the terrestrial planets completely and blow off their primitive atmospheres into space.

In the portion of the solar system from Earth to Jupiter, this process would have stopped before formation of a massive planet like Earth, since we see only one "small" planet (Mars) in this area and thousands of asteroids (whose total mass represents approximately 5% that of the Moon). At the present time, numerical simulation is unable to account for this "failed" coagulation, the causes of which are poorly understood.

One of the most plausible explanations is that Jupiter, formed before the Earth, prevented the coagulation of the small bodies in its immediate vicinity by gravitational slingshot effect. Beyond Neptune, accretion would give rise only to small comets of "dirty" ice. Since accretion processes had slowed down too much, comets could not have reached planetary size, at least not before the young solar wind came along to sweep them up.

This scenario of accretion by giant impacts remains controversial. Nothing about it is certain. Once more, the most plausible theory is not necessarily the most probable. To illustrate the diversity of points of view, we may note that Cameron "coagulates" his planets in a totally different fashion. According to him, turbulence formed rings around the Sun which would break up into half a dozen giant gaseous eddies with rocky cores at least ten times the size of the Earth. The excess gas in the eddies would be blown away by the solar wind. A collision between two of these giants would have shattered their rocky cores, producing the four terrestrial planets. The giant planets would have been born in these local eddies.

METEORITES AND FORMATION OF THE EARTH

The problem of the origin of the solar system has not been definitively elucidated any more than has that of the formation of the Earth. Without the contribution of meteorites of asteroid, lunar, and probably Martian and cometary origin, we would be blind indeed and would be unable to trace its history. We would have no building blocks from which we could put together the Earth by modeling. We would have no idea about the very short time intervals during which accretion processes took place, and hence about the nature of these processes. We would be lacking an essential element of information for understanding how Earth could have heated up, then melted completely.

Researchers have two major difficulties to overcome when they try to retrace the formation of our planet. The first has to do with the fact that the Earth has lost its primitiveness, as the most ancient rocks date back 3.8 billion years,* while we need to travel back further in time, to 4.56 billion years ago. Because of this loss of primitiveness, we are forced to base all our models on the study of meteorites, which are far more ancient. The second difficulty is that when we need "global" analyses, we can only probe the internal structure of the Earth by seismic waves. These measurements, however, reveal only physical characteristics, such as an onion-like layered structure, from which it was deduced that, shortly after the formation of the Earth, its temperature rose suffi-

* For dating age we shall henceforth use a notation different from the ΔT interval used in the preceding chapters. We shall now date back age from the time (T) of the present era, defining T as equal to 0. Thus, an age of 3.8 billion years ago corresponds to the earlier notation of $\Delta T \approx 700$ million years.

ciently to trigger melting of an initial mixture of silicates, sulfides, and metals, producing viscous magmas of different densities. The densest (alloy of iron-nickel and sulfides of these elements) sank to the center of the Earth to form a solid "inner" core with a radius of approximately 1,370 km [850 mi] surrounded by a liquid "outer" core about 2,000 km [1,250 mi] thick. This outer core was surrounded by a rocky mantle 2,900 km [1,800 mi] thick, far less dense, which accounts for two-thirds of the mass of the Earth. At the surface, the unmolten primordial debris became mixed with basaltic lava flows from the mantle to form a very thin superficial crust whose average thickness varies from 6 km [3.5 mi] under the oceans to 40 km [25 mi] under the continents; this even lighter crust currently constitutes the seven major plates supporting the oceans and the continents which float like rafts on the mantle, drifting at speeds of a few centimeters per year. Without the meteorites, our scenario would have gone no further than that!

True, we can make innumerable "in-depth" analyses of Earth samples in the laboratory. But since the crust represents scarcely a thousandth of the Earth's mass, the analysis of this tiny "trace" of Earth provides little useful information for building a formation model, especially since we are really familiar only with samples from the upper crust. As one of my colleagues likes to joke: what hypotheses might a crew of extraterrestrials, landing by chance on Earth and in something of a hurry, formulate on the formation of this planet if they had collected, even for highly sophisticated analyses, only a bucket of sand from Death Valley or a few pounds of lava from Hawaii? The rare earthly samples that are the most interesting must be chosen with the greatest of care; for example, those taken from the few outcroppings of the mantle at the surface of the crust. But even in these favorable cases, the sampling is only of the upper mantle. It is difficult to extrapolate the results of analyses to the entire mantle and more difficult still to extrapolate them to the core of the planet. Meteorites are of great help to us in getting beyond these limitations.

In any scenario of formation of the Earth, its oceans, and its atmosphere, we must first of all choose the starting material. Geochemists are guided by data such as the density of the Earth and the composition of upper mantle outcroppings. Once this first step is accomplished, we must look at the mechanism of accretion and understand how the Earth heated up to melt completely. Then we try to imagine how the oceans and the atmosphere were formed.

The popularity of meteorites, considered as building blocks of which the Earth is made, has to do with their great diversity and particularly the existence in the "differentiated" meteorites of metal phases of iron-nickel, sulfides, and silicates, which we anticipate would be present in the onion-like structure of the terrestrial planets; but what is also interesting about the primitive meteorites is their high concentration of volatile compounds (water, rare gases, organic components, etc.)

which, on degassing, could have generated oceans and atmospheres.

The attempt has been made to "fabricate" the whole Earth with a chondritic material similar to the most primitive meteorites (CI type). The calculations of the geochemists have led to baffling "predictions." If the Earth had formed from this primitive material, which in particular contains far more water in the form of hydrated minerals, and more carbonaceous and nitrogen material than do the other meteorites, then it should be completely covered by oceans, to a depth of at least 100 km [60 mi]! What is more, the iron core responsible for producing the Earth's magnetic field should be larger than the core we know, producing a stronger magnetic field. Another consequence: the quantity of carbon and nitrogen stored on Earth, particularly in the atmosphere, should be at least 10 to 100 times as much as there actually is. Thus, the initial material of the Earth was made of differentiated bodies, having lost a goodly number of volatile elements.

At the present time, scientists hesitate among several scenarios of the formation of the Earth. I will cite only those defended by the modelists I have been talking to (Heinrich Wänke, Ross Taylor, George Wetherill, and Philippe Sarda, of the Allègre group).

According to the first model, the material constituting the Earth accreted in a single stage by coagulation of planetesimals, the largest of which were already differentiated bodies, less rich in volatile elements than the primitive meteorites. Everything the Earth contains, particularly the upper mantle, the iron core, the oceans, and the atmosphere, was produced by melting and degassing of this coagulation of planetesimals. This model is defended by Philippe Sarda, for example.

The second model, proposed by Heinrich Wänke, calls for two different stages during the accretion of planetesimals: two-thirds of the Earth were first formed rapidly, in a few million years, by accretion of a first generation of planetesimals also composed of a differentiated material (enstatite chondrites). Then, during a "late" second stage of accretion, which probably lasted less than a hundred million years, the impact of a second family of planetesimals, richer in volatile elements (comets or C type asteroids, similar to the primitive CI type meteorites), would have contaminated the mantle with siderophile elements (elements preferentially associating with iron) and profoundly modified the composition of the oceans and the atmosphere.

The differences between these two models are not as great as it might appear. First of all, they represent subtleties in the determination and interpretation of chemical and isotopic compositions, such as those obtained for the mantle and the rare gases of the atmosphere. A major difference is their treatment of the formation of the oceans and the atmosphere. In one model, both gas and water are produced by degassing of the Earth after the coagulation of planetesimals. In the other, the impact of cometary or asteroidal bodies, freshly formed and still "a little wet," formed near the frost front and launched by Jupiter's "slingshot" effect, would have dominated the formation of these two ingredients necessary to our life.

But these two models have more points in common than differences. Both are based on direct accretion of *planetary embryos*, during these "worlds in collision," this accretionary tail, whose last footprint is constituted by the great lunar craters. They reject other models such as Cameron's giant gas eddies, or that of far "gentler" gradual accretion caused by infinitely tinier cosmic hailstones regularly raining down.

Indeed, if accretion came about by such a cosmic rainfall, the axes of rotation of the planets would all be perpendicular to the plane of their orbits around the Sun (the ecliptic). This is not the case: all the axes of the planets are tilted at different angles to this plane. Calculations indicate that, to obtain such inclinations, Earth and Venus must have been struck by bodies of the size of Mars and Mercury, respectively. Scientists have looked to these impacts as they inquired either about the origin of the Moon or the very high iron content of Mercury (about 65%). In the case of the Moon, a Mars-type impacter is believed to have ejected a large fraction of the Earth's crust in the form of a layer of debris, which would be low in iron due to formation of the iron core. Re-accretion of this debris, placed in orbit around the Earth, is then thought to have generated the Moon, which is poor in iron like the Earth's mantle and has a very similar isotopic oxygen composition to that of Earth. The impact would have blown off the rocky mantle of Mercury even more completely to produce "our" Mercury, which is 65% dominated by the iron core.

These recent ideas have been "confirmed" independently by numerical simulations. These suggest that about a hundred objects with the diameter of the Moon (about 3,500 km [2,200 mi] across with a mass of 7.35×10^{19} tons), about ten with diameters between those of Mercury (about 5,000 km [3,100 mi]; mass: 3.4×10^{20} tons), and Mars (about 7,000 km [4,300 mi]; mass: 6.4×10^{20} tons), and a few that are more massive than Mars constitute the final population of embryos. This population was essentially "consumed" by the formation of Earth and Venus. Far later, a few smaller survivors formed the accretionary tail, whose effects are observed on the Moon between 4.2 billion years ago and 3.8 billion years ago.

In *one* of these simulations, the "Earth" embryo increased its mass by 50% at the time of the last eight giant impacts, including six bodies with masses between those of Mars and Mercury and two more massive than Mars, with the most cataclysmic impact occurring 11 million years after formation of the solar system. As for Venus, the *same* simulation reveals a simpler "history," dominated by two impacts of bodies with masses intermediate between those of Mercury and Mars, and one impact by a body more massive than Mars.

The embryos, and even certain 10-km [6-mi] planetesimals, were already "differentiated." By comparison with the primitive composition of the CI meteorites, they are poor in volatile components such as water and organic matter. Some of them had already melted.* Their compositions were heterogeneous.

These scenarios, combining observations and numerical simulations, indicate in any event that the primitive history of the Earth was dominated by giant impacts. The variations in density between the planets are no longer related to a hot solar nebula model, in vogue for over twenty years, in which the highest temperatures near the Sun would have favored condensation of denser refractory materials from a gas in the process of cooling. Now the leading theory is one of essentially random impacts by *planetary embryos over 4,000 km [2,500 mi] in diameter*, of heterogeneous composition.

The Earth's landscape about 4.4 billion years ago was that of a molten planet, transformed into gigantic lava flows, subjected to titanic bombardments by bodies over 1,000 km [600 mi] across, surrounded by a highly

* Identification of the energy source that allowed differentiation and melting of planetesimals about 10 km [6 mi] in size, even before they coagulated into far larger embryos, is uncertain. But after the daughter of a radioactive isotope of aluminum (^{26}Al) was found in meteorites, it was supposed that the most effective mechanisms were heating related to decay of aluminum-26. To this was probably added a sort of cooking of planetesimals somewhat similar to the situation in a microwave oven, resulting from the encounter between the magnetic field "anchored" in the solar wind and the planetesimals. This cooking, which depends on the distance from the Sun, must have occurred very early, before the planetesimals were more than 1 km [0.6 mi] across. Then, the gravitational energy released during coagulation of the large planetary embryos was probably sufficient to cause complete melting of the Earth.

unstable atmosphere of changing composition, often blown out into space. It took several hundreds of millions of years before these cataclysms calmed down and the Earth cooled sufficiently to favor the appearance of life.

METEORITES AND THE ORIGIN OF LIFE

For life to appear on Earth, it first needed a primordial soup of simple organic molecules which, by "prebiotic" synthesis, produced more complicated organic molecules considered the building blocks of the living cell, such as amino acids, sugars, nucleotide bases, and lipids. Then something clicked: some of the molecules made from these building blocks began to transfer their information, thus generating the first rough sketch of life. We will not describe this scenario again, but we will point out the role that "exobiologists" assign to meteorites in prebiotic synthesis.

Back in 1953, it all seemed rather simple. The famous Miller-Urey experiments appeared to have solved the problem. A glass retort was filled with a mixture of methane, ammonia, hydrogen, and water, modeling the primitive "reducing" atmosphere of the Earth, like the reducing atmosphere of Jupiter. Then this mixture was subjected to electrical discharges simulating the effect of atmospheric storms, and a primitive soup was produced containing a wide variety of organic molecules, including the famous amino acids from which proteins are built.

Little by little, however, this model of the atmosphere had to be abandoned for converging reasons coming from very different studies. For example, if the Moon had truly been formed when a giant Mars-sized body hit the Earth, calculations indicate that such an impact would have blown the entire primitive atmosphere of the Earth out into space. After this, planetologists concerned with the formation of the Earth were obliged to use a highly differentiated starting material that had lost its volatile compounds such as organic molecules. Finally, atmospheric physicists measured the lifetime of ammonia and methane molecules when exposed to ultraviolet radiation from the Sun (the same radiation that is so dangerous to the skin). Surprise! The lifetimes of these molecules proved to be extremely short—between a week and a few years, meaning that this "sunburn" would have quickly destroyed this type of primitive atmosphere.

Once this reducing atmosphere had been dismissed, we were left with an "oxidizing" atmosphere, of the volcanic type, this time rich in carbon dioxide, nitrogen, and water. It was assumed that this volcanism had been triggered by numerous impacts from the accretionary tail which transformed the emerged continents into a gigantic molten lava flow exhaling huge quantities of these gases. This model appeared to be far more satisfactory because the atmosphere of the Earth then resembled that of its two neighboring planets, Venus and Mars, and not that of the far more massive and distant giant planets as in the preceding model. But an enormous problem arose. When they went back to their flasks and electrical discharges, exobiologists found that this mixture no longer yielded organic molecules! The source of prebiotic molecules on Earth had promptly dried up in their simulations. Now what?

A forgotten hypothesis proposed by the chemist Juan Oró in 1961, at the time the Miller-Urey experiments were going so well, was now revived. Evidence was available at that time that primitive meteorites and comets contained water and organic matter. Oró suggested that the organic matter on Earth could have come from the comets that bombarded it. This "late" accretion of extraterrestrial matter, which probably contributed to generating the oceans, would also have seeded the Earth with prebiotic organic material. The Earth would have to be sufficiently cool for these fragile molecules not to be destroyed by heat. So the start of prebiotic synthesis can be established at approximately 4 billion years ago. The exobiologists tried, unfortunately without much success, to obtain more complex organic molecules, relying on minerals with catalytic properties such as clays.

I agree with André Brack, a chemist by training and an exobiologist by passion, that it is unlikely the comets survived their impact with the Earth. On the Moon, where meteorites are not slowed down by an atmosphere, no meteorite (large or small), and certainly no trace of the constituent organic molecules of primitive meteorites is found.* The only trace of meteorites in the lunar soil is a scattering of precious metals such as gold, platinum, and iridium released when meteorites and micrometeorites volatilized as they hit the Moon. But the comets in Oró's model appear to be too large to have slowed down in the Earth's atmosphere, so they must have exploded cataclysmically on the ground, as did those that struck the Moon. So it is unlikely that organic molecules could have survived such explosions intact.

On the other hand, micrometeorites survive far better than predicted on impact with the atmosphere, and they contain organic matter. Moreover, they represent by far the major source of extraterrestrial material on the Earth, if we exclude bodies over ten meters [33 feet] in size which explode because they are too massive to be slowed down by the atmosphere. Hence we proposed the idea that micrometeorites, rich in organic matter, could have supplied the material necessary for prebiotic synthesis.

We will go further. Each micrometeorite could have functioned like a microfactory of organic molecules, called by Brack "chondritic microreactor." In fact, micrometeorites contain all the ingredients whose existence was postulated by exobiologists in the previous models: not only organic matter but also clays and numerous catalysts necessary for accelerating reactions

* Small organic molecules have been found, such as methane (CH_4), which was synthesized when the solar wind (bringing C and H) was implanted into the lunar grains to very shallow depths.

with water, such as oxides and sulfides of iron, nickel, and chromium, and metallic chunks of iron-nickel. All these reagents are confined in the tiny volume of the micrometeorites. By contrast, other models have to tackle the severe problem of the dilution of chemical species in water. So we would assume that in the numerous pools of highly mineralized water that existed about four billion years ago (when the flux of extraterrestrial material impacting the Earth was at least 1,000 times what it is today), micrometeorites underwent a chemical reaction of "hydrolysis" with water, during which the building blocks of the living cell were synthesized.

Our small international team of Americans, Austrians, French, and Japanese, is working on this theory. An alternate theory, proposed a decade ago, placed this prebiotic synthesis on the ocean floors in highly mineralized hot hydrothermal springs. These springs were discovered in 1977, at a depth of 2,600 m [8,500 ft], not far from the Galapagos Islands, by the oceanographic research submarine *Alvin*. There was an extraordinary profusion of life, over areas of several hundred square feet, including heaps of 6-ft-long worms, giant mussels, white crabs, and other forms of life. These geysers of life have developed away from sunlight, without oxygen, and at a high temperature!

The springs are located along the submarine mountain chains (the "backbones" of the ocean floors). They result from seawater which has infiltrated to great depths through microcracks in the rocks, reaching temperatures of over 300°C [570°F]* (without boiling because of the enormous pressures achieved at these depths) and reaching high concentrations of minerals from the rocks. The water returns back to the surface by either seeping slowly or shooting up as fast as over 3 m [10 ft] per second in the form of geysers. Many of these geysers look "black" from being laden with metallic sulfides that precipitate in the form of very fine particles as soon as they come into contact with cold water, giving them the black appearance that leads to their name of black "smokers." In this model, the carbon dioxide, nitrogen, and hydrogen** thus released at the sea floors would react to synthesize organic molecules.

A weakness of this model is that the temperatures are far too high in the hydrothermal springs to allow the survival of complex and fragile organic molecules.

These two competing models (micrometeorites and hydrothermal springs) have to be compared in the laboratory, using difficult simulation experiments whose results will always be somewhat uncertain. For example, the reaction rates are so slow that they must be accelerated if reaction products have to be seen in a human lifetime; but then the reaction sequence might be different from that which occurred in nature. One member of our team, the American Everett Shock, is working on both types of synthesis. We will in a few years' time be

* This heating is due simply to the thermal gradient of the Earth which increases as a function of depth.

**The hydrogen results from decomposition of water when it reacts with certain minerals formed in the hydrothermal springs.

in a good position to determine which one of these models was responsible for our own existence.

METEORITES AND EVOLUTION OF LIFE

The Earth can be regarded as a target on a gigantic, cosmic firing range. The shooting has left behind enormous scars (impact craters) on planets such as Mars and Mercury and on numerous planetary satellites, including the Moon. On the Earth itself, despite the forces of erosion and plate tectonics which remodel the relief, we can still count nearly 130 impact craters. They may be as large as 200 km [125 mi] across and their ages go back from recent times (30,000 years ago for the Barringer Crater) to about 1.8 billion years ago in the case of the Sudbury crater in Canada.

If we keep in mind the gigantic explosive power of the massive bodies that strike it (about 5 billion Hiroshima-type atomic bombs for a 10-km [6-mi] body), we have every reason to believe that certain living species must have gone through the effects of cataclysmic impacts on Earth. Since such impacts are inevitable in the future, it is important to understand their effects. It makes sense to compare the dates of major extinctions of species to those of the great impact craters still visible on Earth. Do the dates of crater formation match up with the triggering of any extinctions? It also seems important to organize a sort of telescopic "watch" as an early warning system for the arrival of killer bodies on Earth.

These ideas were foreshadowed about fifty years ago by a few visionaries such as F. Watson and Ralph Baldwin, a "lover" of the Moon, and later by Eugene Shoemaker. But they fell into oblivion. An article published in 1980 by the physicists Alvarez (father and son), the importance of which was termed "cosmic" by the great paleontologist David M. Raup, was a rude awakening for us. For over ten years scientists from very different disciplines have been working together on the various scenarios of "mass extinction" that are believed to have affected the Earth globally.

Specialists chart the history of our planet since 4.56 billion years ago on a time scale whose intervals become increasingly small as it approaches present time. There are six major periods covering fifteen principal periods of which the most recent (between 205 and 1.5 million years ago) are themselves divided into fifteen periods whose names are surprising to the newcomer: Ordovician, Silurian, Dogger, Cenomanian, Pliocene, and so on. The dates separating the periods that extend over the last 300 million years frequently correspond to mass extinctions, when numerous living species disappeared simultaneously and were followed by rapid evolution of the survivors. In the last 300 million years there have been about twelve such crises, the most massive occurring in the Permian (about 230 million years ago), a period during which nearly 90% of living species disappeared.

These extinction crises, called "decimations" by the Harvard paleontologist Stephen J. Gould, are superimposed on a sort of "background noise" attributable to normal "daily" extinctions. Even before the appearance of humans, this background noise existed, of course. The accumulation of these "slight and massive" extinctions is impressive: each species living today corresponds to some hundred species that have disappeared. At the present time, the extinction background noise is increasing considerably because of the proliferation of human activities, which have become the greatest killers of all time, causing thousands of living species to disappear every year. Still, we must remember that there *are* at least several million of them.

In making these fascinating studies, researchers must use fossils that have been well preserved in a regular stack of strata of sediments susceptible to precise dating. This means a choice of good sediments that are not too old. So researchers use marine sediments less than 300 million years old,* which contain the most abundant terrestrial fossils, those of small marine organisms whose shells are about the size of a grain of sand —such as the foraminifera classified in several thousand varieties. These fossils are deposited in strata of sediments of the ocean floors. Tectonic movements can stir them up to the surface, yielding easily accessible sedi-

* The fossils deposited before this date are very poorly preserved, as their carbonate constituents decay and tectonic activity recycles about one-quarter of the ocean bed approximately every 60 million years.

mentary strata, such as those outcropping in the gorges near the medieval town of Gubbio in northern Italy. Massive extinctions are manifested by a sudden drop in number of these small organisms in one of the layers. Generally, these extinctions also involve many other species, but on a lesser scale.

The best known of them is without a doubt the famous "K-T" (the abbreviation used for "Cretaceous-Tertiary") boundary, dated 65 million years ago. This extinction represents a decisive stage in the evolution of life, during which not only did nearly 50% of small marine organisms disappear, but the dinosaurs, who had reigned over the Earth for the previous 150 million years, also perished. They were replaced in the vacant ecological niches by a profusion of mammals that are our ancestors.

What is the origin of these gigantic upheavals? The case of the dinosaurs attracted so much interest that in 1980 one could count nearly a hundred different explanations for their disappearance. The more serious ones can be related to two principal and conflicting scenarios: a "gradual" extinction that took several million years and a far more sudden or "catastrophic" extinction, occurring in less than 100,000 years. The imagination of investigators was unleashed. Some of the invoked causes are: explosion of a star near the Sun (supernova), causing acid rain that destroyed plant life; a genetic defect blinding these animals with cataracts; another genetic defect that afflicted them with slipped disks and backache, particularly handicapping for these mastodons; skin diseases; a mouselike shrew that developed a special fondness for dinosaur eggs; an irresistible and uncontrollable drive to collective suicide of the type that today leads to the beaching of certain schools of whales and dolphins; a thickening of the shells of their eggs, preventing them from hatching; and so forth.

The proponents of the sudden extinction theory, wonder whether it was caused by volcanic cataclysms altering the climate that could recur in the future, or by an as yet mysterious astronomical phenomenon that triggered broad climatic changes, in the manner in which a tiny wobble in the Earth's rotational axis (due to the gravitational pull of other planets*) brought about periods of glaciation (ice ages) estimated to be about 100,000 years apart. People worry whether such catastrophes could affect one particularly important living species: *Homo sapiens*.

But in 1980, the Alvarezes described in the prestigious journal *Science* the discovery, in the sedimentary formation of Gubbio, of a very high concentration of iridium in a clay layer dating back 65 million years, a few centimeters thick, light-colored, squeezed in between far darker sediments. This result and its interpretation set off a revolution whose effects are still felt today.

Luis Alvarez received the Nobel Prize in physics in 1968 for developing new instrumentation highly useful

---

* Like a spinning top, the axis of rotation of the Earth undergoes a slow precessional motion about itself that is responsible for the ice ages.

in nuclear physics. In 1970 he began his collaboration with his son Walter, a young and still relatively unknown geologist who was teaching at the University of California in Berkeley. Their project was to study extremely rare elements in terrestrial rocks such as iridium, then look at ancient sediments with the hope of being able to determine their rate of deposition. The attention of the two researchers was understandably directed at sediments at the K-T boundary.

In fact, they knew, because of *the existence of meteorites* and *measurements by meteoriticists*, that iridium is about 20,000 times more abundant in meteorites* than in the rocks of the Earth's crust. Thus they expected to measure very low iridium concentrations in the sediments, reflecting the gradual process of contamination by the rain of meteorites, assumed to be constant. The thicker the layer of sediment deposited in a year (that is, the higher the rate of sedimentation), the more iridium, they reasoned, would be diluted in the sediments.

In the Gubbio clay layer, they found that the iridium concentration jumped suddenly to a value 300 times higher than in the rock surrounding the clay layer. This stupefying discovery led them to investigate this iridium layer all around the globe. Thus it was shown that the rain of iridium is truly global, as it is found at over seventy-five sites scattered throughout the world, covering both

* Since 1986, the remarkable work by the Frenchman Robert Rocchia and his co-workers relating to detection of cataclysmic impacts of the past has shown that micrometeorites contain high iridium concentrations similar to those of meteorites.

sea floors and continents. How is one to account for this sudden dusting of iridium at the very time the dinosaurs were disappearing? The two Alvarezes advanced the idea that the iridium came from the impact on Earth of a 10-km [6-mi]-sized body, exploding like 5 billion Hiroshima-type atomic bombs, capable of producing a crater between 150 and 200 km [95 and 125 mi] in diameter and 10 to 20 km [6–12 mi] deep. This body would have thrown up thousands of billions of tons of dust into the atmosphere.

This article itself went off like a bomb in *Science*, taking scientists by surprise, but the counterattack was soon launched and critiques of the model flooded in. On Earth, a few large impact craters several hundred millions of years old are observed. Beginning in 1981, the objection was raised that some trace of the famous K-T crater should have been found, while no such crater was reported at that time. The Alvarez contingent replied that it could lie beneath the oceans. But the many cartographic surveys of the ocean floor should have detected it. Reply: approximately one-quarter of the ocean floor is renewed every 60 million years or so, "digested" by the Earth's mantle in subduction zones. The crater could have disappeared, perhaps after triggering its own mini-subduction zone.

The paleontologists, somewhat taken aback by this sudden intrusion of physicists into their tradition and their routines, then pointed out that the twelve extinctions known as of this date had been far slower than the expected effects of a cataclysmic explosion. The duration of the extinctions, which are difficult to measure exactly, always exceeded a million years and the dinosaurs themselves started to die out 10 million years before the presumed date of the impact.

But in fact, the paleontologists have not yet succeeded in agreeing among themselves. In 1991, the team of P. M. Sheehan, after lengthy and meticulous field work in the American West, clearly showed that the diversity of the dinosaurs did not decrease gradually over a period of 10 million years, and that their extinction was indeed sudden. But the most "gradualist" diehards stuck by their positions, even though they began to admit the reality of the explosion, conceding only that it may have given the coup de grâce to a population that had already begun to dwindle under the effect of an ecological stress, probably linked to a major drop in sea level of about 150 m [500 ft].

At the present time, paleontologists, meteoriticists, astronomers, astrophysicists, volcanologists, and "impacters" (specialists in impact craters on planets), forgetting their quarrels, are actually working together. The problem is too difficult and too important for any one discipline to tackle alone.

To explain this phenomenon which marked the K-T boundary, the hypothesis of an enormous impact still seems the most plausible today. The "gradualist" theory of a giant volcanic eruption, like that of Deccan in India, in vogue for several years, has just been seriously shaken by recent studies presented in June 1992. These studies demonstrate, through precise dating, that none of the major extinctions observed in the last 300 million years or so can be attributed to such volcanic activity.* Another blow to this theory: Deccan's lava contains too little iridium to account for the quantity of this element trapped in the K-T boundary sediment layer found all around the globe.

As to the existence of the crater, its probable position has been deduced by following the direction of *increasing* iridium concentrations which converge at point zero, located in North America (and not in the Deccan plateau). It is believed that not just one, but two craters of the right age have been found, which suggests that the collisions might have been close together in time: the Manson Crater (in Iowa) with a diameter of 36 km [22 mi], dated back 65 million years, and the Chicxulub Crater (in Yucatan, Mexico), 200 km [125 mi] across, which is the right size for being caused by a 10-km [6-mi] body but whose dating still needs to be improved, although we have a range of between 60 and 80 million years ago. A correlation has been established between this rise in iridium and an increase in both sediment grain size and abundance of minerals "shocked" by the intensity of the explosion. These last two features are verified in the "ejecta blankets" of craters observed either on the Moon or with artificial explosions.

---

* During this episode an enormous flow of lava which covered an area one half the size of Texas gave off volcanic fumes for over a million years.

Numerous problems remain unsolved. Why, for example, do sediments of other extinctions show no net iridium enrichment? Why is it that side by side with the victims (dinosaurs, all marine reptiles, most marine shellfish, most marine plankton, and flying reptiles), there were also survivors whose populations were not substantially affected, such as tortoises, mammals, lizards, snakes, crocodiles, and birds? In short, why did the impact kill so selectively?

Climatologists have presented a catastrophic scenario where the effect of the impact is treated like a "nuclear winter," worse than the winter expected from the explosion of the world's entire stockpile of nuclear weapons, which represent a threat of approximately twenty tons of dynamite for each inhabitant of the United States and the former Soviet Union. A body 10 km [6 mi] in size, as we have said, releases an explosive power of *5 billion atomic bombs of the Hiroshima type*. Titanic fires would have flamed up over the continents. The gigantic quantity of gas and dust injected into the atmosphere as well as the fumes, would have caused darkness to descend on Earth, lasting for months. The production of nitric acid from the burnup of atmospheric nitrogen would be colossal, acidifying not only the soil but also still and running water, and most probably the surface of the oceans as well. Months after the dust had cleared from the atmosphere and daylight had dawned once more, the enormous quantity of water vapor and carbon dioxide injected into the upper atmosphere could produce a greenhouse effect, pushing the temperature up suddenly 10°C [18°F]. This would make things worse still!

In fact, a layer of ash has been found in the K-T layer whose carbon has an isotopic composition suggesting that it derives from the burning of living plants. The quantity of this ash suggests that nearly 80% of the world's plant life—which was far more abundant than it is today—would have burned in this cataclysmic fire, on a planetwide scale. The scenario of a titanic nuclear winter thus becomes more and more probable.

Before the work of P. M. Sheehan's team, which needs to be verified at other sites scattered through the world, astronomers felt obliged to prolong the duration of the effects of the explosion by a mechanism that was able to trigger "bursts" of cosmic bodies in rapid succession (less than a million years apart), producing a cascade of nuclear winters rather close together in time. Today it seems easier to "fire off" such bursts with comets rather than with asteroids. Various mechanisms have been studied, all of which have in common that they present "insurmountable difficulties" in the words of their opponents.

According to this cometary hypothesis, which is my favorite, the "magazine" would be Oort's cloud itself, containing over one trillion rounds of ammunition. The trigger would once again be the gravitational force released when the Sun, in its galactic motion, approached a sufficiently massive object. The list of these objects is "exotic." It includes both a far-off planet and a discrete companion of the Sun (whose orbit would be right in Oort's cloud*), as well as the interstellar clouds that the Sun encounters as it leaves and reenters "perpendicularly" the galactic plane (where these objects are concentrated) with an apparent periodicity of about 30 million years.

However, when we compare the dates of the extinctions with the dates of impact craters on Earth, the existence of such periodicity is not actually borne out, while indicating that a cataclysmic impact of the K-T type should occur roughly every 30 million years. The inaccuracy of the dating and the rarity of the events actually make any conclusion uncertain. However, some crater ages seem to cluster, which would support the existence of bursts of several impacting bodies.

After observing innumerable craters on many solar system objects and making these detailed predictions, numerous physicists, astronomers, astrophysicists, meteoriticists, planetologists, and climatologists continue to attribute a major role in mass extinctions to impacts, even though a goodly majority of paleontologists remain on the sidelines or declare themselves frankly hostile to a scenario of this type.

Although doubts may arise about any long-term prediction, it is fashionable to prophesy major future extinction crises. Various possible end-of-the-world scenarios have been proposed in recent years. The French astrophysicists Jean Audouze and Hubert Reeves, and, in

* These objects are invisible through ordinary telescopes but the IRAS infrared astronomy satellite should have been able to detect them.

the United States, Carl Sagan have often told us how the Sun, when it becomes a red giant in 5 billion years, will swell up to several thousand times its present size to the point that its atmosphere incinerates the Earth. At the present time, many "environmental institutes" are being overhastily created throughout the world in an attempt to predict whether human activity will lead to an irreversible greenhouse effect, producing major climate changes. The number of expected deaths due to skin cancer if the ozone hole in the atmosphere gets any larger has already been calculated. Others invoke the appearance of a virus even more terrible than the AIDS virus. If the pessimistic predictions of Jacques Cousteau become reality, humankind, due to galloping demographics and world hunger, will itself set off the most massive of all extinctions: the rich countries will use apocalyptic weapons to thrust back the assault of the starving billions of the Third World who have nothing to lose.

We should also be concerned, without giving way to panic, about the devastating effect of a cataclysmic cosmic impact. On March 31, 1992, NASA presented to the United States Congress a summary of a study it had commissioned from two groups of top scientists. The first group was asked to establish the mass threshold for a truly global catastrophe corresponding to impacting bodies 1 to 2 km [0.6 to 1.2 mi] in size, and to determine their trajectories when they came close to Earth. The second group, meeting at the great Los Alamos National Laboratory in New Mexico (where the first atomic bomb was developed), was charged with proposing methods to intercept these intruders in sufficient time and knock them off their collision course with Earth, if necessary by a few well-placed H bombs. It is curious that, in this sort of "star wars," humankind would owe its survival to H bombs and guided missiles hitting the target with an accuracy of about a hundred feet.

Discoveries of such bodies have been coming in thick and fast ever since the geologist Eugene Shoemaker and his wife Carolyn were converted back, some ten years ago, to careers as astronomers and became the greatest hunters of killer asteroids visiting the near environs of the Earth. For the only family of asteroids whose orbits cross (or will in the near future cross) the orbits of the Earth or Mars (Apollo, Amor, and Aitken), one object 1 km [0.6 mi] in diameter passes about every twenty years *between the Earth and the Moon.* They are supposed to be over two thousand in number and a quarter of them would end their interplanetary wanderings by striking the Earth, with an average interval of a few hundred thousand years between two impacts. To date, only 10% of these objects have been counted. With today's telescopes, it would take about twenty-five years of uninterrupted observations to make a complete survey. The smaller objects would not cause a global catastrophe; they would present a real danger only if they impacted a densely populated area. In that case, an object the size of a 600-foot steamboat, exploding like a thousand Hiroshima bombs, could cause tens of millions of deaths. But such objects are expected to strike only every 5,000 years. However, it is very difficult to observe them and they are very large in number. With the resources we have, it would take several centuries before they could all be pinpointed with sufficient accuracy to predict their effects.

Long-period comets that pass near the Sun for the first time are very beautiful, with their long tails of gas and dust trailing over hundreds of millions of kilometers. But they seem to be far more treacherous than the asteroids, because they can be detected only a few months before they strike the Earth, while the fall of an asteroid can be predicted some ten years in advance. This leaves very little time to organize a response. What is more, they just may arrive in "bursts"! Happily, it is estimated that the probability of their hitting the Earth is one-tenth of that calculated for asteroids.

We have no reason to believe that the apocalypse will be tomorrow, but it seems desirable to organize a telescopic "watch" at the European level to complement the U.S. watch. For example, the efforts of the French astronomer Alain Maury (Nice Observatory) could be supported, as could be the attempts to set up an automatic telescope to detect killer bodies at high latitudes poorly covered by other observatories, such as those in Spitsbergen (in the Arctic) and Antarctica. What would happen to our climate and our coastal populations if a killer body were to strike Antarctica where nearly 90% of the Earth's ice is stored?

CONCLUSIONS

Meteorites of all sizes are unequalled archivists of the primitive history of the solar system. Hunting for them, which is not about to stop, will bring new surprises and reveal new solar system objects. Unknown types of meteorites, probably collected in Antarctica and the hot deserts, will one day steal the starring role from the primitive meteorites. Within the next decade, more powerful microanalytical instruments operating automatically will be able to analyze meteorites in extraordinary detail, discovering, in particular, new presolar interstellar grains. With these new techniques, scientists will be able to analyze the first samples taken by automatic space probes from the very surface of comets or asteroids.

These new types of analysis, together with progress in nucleosynthesis and the astronomy and astrophysics of molecular clouds and protostars, will allow us to trace probable (not merely plausible) scenarios of the origin and evolution of our most primitive mineral ancestors. We will also understand far better the effects of the early "worlds in collision" on the formation of planets and smaller bodies of the solar system.

Every year, comets and asteroids send to Earth nearly 20,000 tons of micrometeorites and less than 100 tons of conventional meteorites. Micrometeorites may have triggered the synthesis of prebiotic molecules on Earth, to which we owe the extraordinary appearance of life in an obscure, hot pool four billion years ago. The existence of meteorites appears to be a miracle, probably linked to the rapid formation of Jupiter—which prevented their parent bodies, the asteroids, from coalescing into a small planet orbiting between Mars and Jupiter.

It is a strange paradox that this same miracle, both beneficial for our knowledge and decisive for the origin of our life, bears within it the germ of mass extinctions from cataclysmic impacts between asteroids and the Earth. The bearer of life, it may also be the herald of our death. Unlike the dinosaurs, however, humans should be capable of predicting and therefore preventing such impacts, unless problems of overpopulation, famine, or uncontrollable acts of madness by fanatics push the Earth's people into using their new power of "self-destruction"—even before the next cataclysmic impact.

BIBLIOGRAPHY

ALLÈGRE, C., *Introduction à une histoire naturelle* [Introduction to a natural history], Paris, Fayard, 1992.

BRACK, A. and RAULIN, F., *L'Evolution chimique et les Origines de la vie*, [Chemical evolution and the origins of life], Paris, Masson, 1991.

BROWNLEE, D. E., "Cosmic Dust: Collection and Research," in *Annual Review of Earth and Planetary Sciences*, vol. 13, p. 147, 1985.

BURKE, J. G., *Cosmic Debris, Meteorites in History*, Berkeley, University of California Press, 1986.

CAMERON, A. G. W., "Nucleosynthesis and Star Formation," in *Protostars and Planets* III, ed. Levy, E. H., Lunine, J. I., and Matthews, M. S., University of Arizona Press, in press.

Id., "Origin of the Solar System," in *Annual Review of Astronomy and Astrophysics*, vol. 26, p. 441, 1988.

CHAPMAN, C. R. and MORRISON, D., *Cosmic Catastrophes*, New York, Plenum Press, 1989.

KERRIDGE, J. F. and MATTHEWS, M. S., ed., *Meteorites and the Early Solar System,* Tucson, University of Arizona Press, 1988. (This compilation, which includes 51 high-level contributions, is the bible of meteoriticists.)

LEMAIRE, T. R., *Stones from the Stars. The Unsolved Mysteries of Meteorites*, Prentice Hall, 1980.

MCSWEEN, H. Y., Jr., *Meteorites and their Parent Planets*, Cambridge University Press, 1989.

TAYLOR, R. S., "Accretion in the inner Nebula: the Relationship between terrestrial planetary Compositions and Meteorites," in *Meteoritics*, vol. 26, p. 267, 1991.

WETHERILL, G., "Formation of the Earth," in *Annual Review of Earth and Planetary Sciences*, vol. 18, p. 205, 1990.